LIFEBOAT CITIES

BRENDAN GLEESON

UNSW
PRESS

A UNSW Press book

Published by
University of New South Wales Press Ltd
University of New South Wales
Sydney NSW 2052
AUSTRALIA
www.unswpress.com.au

© Brendan Gleeson 2010
First published 2010

10 9 8 7 6 5 4 3 2 1

National Library of Australia
Cataloguing-in-Publication entry
      Author: Gleeson, Brendan, 1964–
      Title: Lifeboat cities/Brendan Gleeson
      ISBN: 978 1 74223 124 2 (pbk.)
      Notes: Includes index.
         Bibliography.
      Subjects: Social evolution.
        Social change – Australia
        Urban policy – Australia.
        Australia – Social conditions – 21st century.
      Dewey Number: 307.160994

Design Avril Makula
Cover Nada Backovic Designs
Cover images: iStock
Printer Ligare

This book is printed on paper using fibre supplied from plantation or
sustainably managed forests.

LIFEBOAT CITIES

PROFESSOR BRENDAN GLEESON is Director of the Urban Research Program at Griffith University. He is the author of *Australian Heartlands: Making Space for Hope in the Australian Suburbs* (winner of the inaugural John Iremonger prize in 2006) and co-author of *The Green City: Sustainable Homes, Sustainable Suburbs* (2005). His 'Waking from the Dream' essay, published in the *Griffith Review* in 2008, was included in *The Best Australian Political Writing 2009*, edited by Eric Beecher.

CONTENTS

For Ulrike, Julian and Alison

1

A NEW AUSTRALIA HERE

A CENTURY OR MORE AGO, in the wake of a grinding drought and prolonged recession, a group of Australian dreamers left to found a 'New Australia' in the jungles of Paraguay. The early promise of *Australia Felix* had failed them: it was time to move to a different New World, where the ideal of egalitarianism could be forged again. The experiment failed in very Australian ways: squabbles about grog and work created searing tensions. The dream of something different faded very quickly.

Meanwhile, back in Australia, pragmatic reformers were pushing the claims of fairness in less spectacular, but quietly effective ways. A political party representing the working class was formed, and, over time, the bosses were forced to guarantee the conditions of a decent, if basic, life for ordinary Australians. This new dispensation – the Australian Settlement – fell short of the egalitarian utopia that the dreamers sought in Paraguay. It describes a long, roughhouse period of conflict and compromise between capital and labour, during which many gains for working people were won. A state sector emerged to support the needy and the infirm. Its institutional forms were in some instances monstrously callous. Slowly and surely, however, social support was improved and extended in ways that would have seemed unimaginably generous to the poor of the 19th century.

The Settlement was far from perfect, but it was a good deal better than the brutish laissez-faire society of the 19th century. This was a time when economic liberalism and class prejudice defined politics and power in England and its colonies. The model failed miserably and was rejected, only to return a century later with the rise of 'neoliberalism' from the 1970s. By then, the mass consciousness had largely forgotten how vicious a society founded on extreme liberalism could be.

The Australian Settlement was a deliberate attempt to steer us away from the depression and conflict generated by the laissez-faire model of the second half of the 19th century. It set the conditions for a long period of peace, prosperity and (relative) fairness in the century we have just left. Spectacular interruptions – two world wars and a global depression – were followed by new bursts of zeal to re-establish the Australian Settlement and improve its basic features.

There were two great omissions from the project: an unwillingness to acknowledge, let alone appreciate, the original inhabitants of 'terra nullius' and a failure to comprehend the land that they had carefully nurtured for millennia. It was a failure to understand our own nature. And it cost us dearly. We proved poor stewards of the land. We were insensitive, sometimes aggressively, to cultures we regarded as different or whose claims were inconvenient to us.

The reformism rejected by the Paraguayan exiles left us vulnerable to the return of failed ideas, namely laissez-faire. When the developed world's economy ran out of steam in the early 1970s, the sacred keepers of this abortive 19th-century ideology sought its reinstatement under the banner of 'neoliberalism'. By this I mean the pro-market mindset that began to grip public and political culture in Australia, but for which, paradoxically, there has never been popular enthusiasm. Markets came to be seen as good ends in themselves rather than one means by which society could reach the goal of collective welfare. This conviction carried the political projects of Thatcher and Reagan through the 1980s, and also animated the reform agendas of Hawke, Keating and Howard in Australia. Neoliberalism was largely confined to Britain, the United States, New Zealand and Australia (Australian sociologist Michael Pusey has termed it the 'English-speaking disease'). It took different forms in each of these nations and had quite different impacts on their political cultures and institutional arrangements.

In Australia, we did not, thankfully, dismantle the welfare state in the way that Britain's Thatcher did. And Reagan's callousness towards the poor was not deemed acceptable here. What united these various political reform projects was their rather similar impact on the political imagination, which closed progressively to the point where in 1992 one US neoconservative, Francis Fukuyama, declared the 'end of ideology'.

Australia bore many of the material consequences of the neo-liberal agenda, through privatisation, deregulation and corporatisation. And yet I think it was this closing of political and institutional minds to the possibilities of collective effort that most damaged our prospects. We have now have generations of policymakers and public officials who simply have no idea how to design and implement collective solutions to the increasing array of problems thrown up by deregulated markets. The many struggles and setbacks that have attended the Rudd government's nation-building programs are partly attributable to this dimming of bureaucratic skills and mindsets.

Many will condemn me for these observations. They'll point to the superficial affluence of the aspirational economy that thrived in the late 1990s and into the new millennium before hitting the wall of the global economic crisis in 2008. They will ignore the relentless depletion and despoliation of our landscape and national institutions and the growing cleavages between the winners and losers in the great game of reform. They will perhaps not even be aware of the withering of ambition and capacity in our political and institutional cultures.

Neoliberalism is no longer the issue, though its legacies will continue to haunt us. It's what must follow this increasingly dysfunctional project that must now concern us. The right-wing ideologues in the universities and think-tanks who carried the flags for neoliberalism have been marginalised to some extent. Witness the big-spending, big-taxing pragmatism of the Howard years. It must have driven them mad. Two dangerous tendencies seem to be marking our way forward now: political pragmatism (survival whatever the cost) and scepticism (an aggressive distaste for new thinking, especially anything that challenges the market status quo). These maladies thrive because public culture has been weakened by two linked processes, both encouraged and abetted by a succession of reform-minded governments since the early 1980s: a narrowing of civic discussion and a relentless concentration of media power (the latter driven more perhaps by favours than by ideology). Civic debate and possibility have been narrowed by many so-called reforms, especially the corporatisation of the universities and the public service. These trends have reinforced a view among elites that market power, not civic politics, is the true arbiter of democracy. Michael Pusey writes:

> The market was meant to bury deliberative politics, to reduce popular expectations of government, to redefine politics as economic management *tout court* and to neutralise normative culture. To use Francis Fukuyama's phrase, it was meant to

> bring us to the end of history and even to kill the shaping
> influences of memory and history in national politics.[1]

Our straitened political culture quickly dismisses any thinking that strays outside the wilting ambitions of liberal democracy. Its concept of civic freedom and human realisation is remarkably unquestioned in popular or political culture. It will be questioned in this book.

We slaves to consumerism, to corporate power, to heartless technologies, have been encouraged to accept a rather miserable sense of liberty. Worse, we've neglected the extent to which neoliberalism has been a self-serving project effecting massive wealth transfers that have impoverished the vulnerable and robbed much of the life and purpose from our civic institutions and culture. Neoliberalism's attack on the civic realm is what I have elsewhere termed *The War on Terra Publica*.[2] Public atrophy has been mirrored by social division. Decades of grinding reform have piled increasing riches on an ever-narrowing social base in Western countries. And the neoliberal project has not restricted itself to Western countries; it has fostered elitism and polarisation in other parts of the world as well.

There have been repeated attempts during the Long March years of the 1980s and 1990s to generate alternative thinking and to expand our understanding of freedom. In a climate of resource-starved and diminished public discussion, most of these attempts withered away quickly, but some ideas and their proponents were not smothered by the sandrifts of cultural indifference. Australian thinkers such as Frank Stilwell, Eva Cox, Boris Frankel, Leonie Sandercock and Patrick Troy have repeatedly called conventional wisdom to account. The efforts of the stout kept the flame of alternative aspiration burning.

The mantle of indifference is at last falling away – indeed, being torn aside – by growing recognition of the structural failures of neoliberalism, and of its pet project, globalisation. From 2007, the whole project began to implode rapidly, and for many shockingly. The Global Financial Crisis (GFC) knee-capped the West and helped to effect the previously unimaginable in America: the election of a black President. There wasn't much time for celebration of this turning point, as he and the other guardians of the global economy were immersed in an economic and social emergency. By late 2009, many wanted to believe that the worst had passed; certainly in Australia, which has navigated the storm much better than most. And yet the terrible social cost of the meltdown cannot be discounted and continues to accumulate, especially in the United States and in the most vulnerable parts of the developing world.

The future is hardly certain, but neoliberalism has certainly been discredited, at least for now. At the same time, the world 'community' (such as it is) is struggling to comprehend and acknowledge an epic shift in the global climate that potentially threatens the viability of our species, and many others besides. We remain mired in the work of comprehension and recognition, still a long way from any agreed response to our greatest act of folly, global warming. Confounding the crises, and making them more volatile and unknowable, is our rapid descent into resource shortages, especially of food, water and that blessed pestilence, oil.

This great unravelling echoes the multiple defaulting of laissez-faire in the late 19th century. In Australia, and Europe, this generated a ferment of new thinking. Some sought a New Australia elsewhere, others strived to reform and make safe the land on which they stood. Today we find ourselves again exhausted and depleted by natural and economic disruption. Worse, we face these familiar threats at greater, even more frightening scales. The spectres of climate change, resource insecurity and global economic recession loom over Australia.

This may be all very well, at least for those who seek a radical departure from the stultifying, unjust rule of neoliberalism. But what worries me most is the possibility that the long blanketing of our political culture has suffocated and extinguished its life-force. I cast this long night under the blanket rule of neoliberalism as a collective dream. It was a time when our species' aspirations, and, frighteningly, our survival instincts, were anaesthetised by an ideology that asked us to refuse nature. We were denaturalised in two ways: first, by refusing our own nature – that is, human interdependency; and second, by repudiating our place within nature – that is, our natural dependency. This was always going to end in tears. The progressive breakdown of human solidarity in the past few decades has produced widening wealth disparities, rising cultural tensions, and a seemingly endless war between the West and its discontented rivals. The great natural ruptures that have occurred need no further underlining.

If we awake without life-force, entirely denuded of political and cultural imagination, in a kind of suspended animation, we risk falling prey to some new project of power. Our great species' potential for collective fulfilment that stirred into life with modernity will remain stalled, or redirected to nefarious ends. We will follow the leaders again, and won't achieve the new forms of freedom and fulfilment that flourish when our species acts collectively, with humility and in concert with sustainability.

My worry is that the political dream of neoliberalism will be replaced by another dream, a new form of human braggadocio that will march us finally over the cliff. What is this new arrogance that might emerge at a time when circumspection and humility are called for? I think it's 'scepticism' of the sort that has aggressively asserted itself in Western public cultures in the last couple of years, especially in climate debates. Its proponents invoke 'science' as a cloak for their corrosive doubt. Their mischief protects the status quo and the forces that are driving us towards the failing of nature. In doing so, they twist and misrepresent the notion of scientific scepticism. Doubt is a moment, a point of tension in scientific thinking, counterposed to Reason. Doubt and Reason are the mutually restraining twins bequeathed to our thought by the Enlightenment. Modern science is not to be reduced to either. The contemporary sceptics are reductionists, not scientists, and have a mindset more like the one that prevailed in the half-light of the pre-modern world than the Enlightenment way. You wouldn't want to fly in an aeroplane designed by one of these sceptics. We should not let them get their hands on our scientific or political institutions. They are, put simply, beyond reason.

I'm not entirely down on collective dreams. Only the ones that imagine us liberated from, or in simple command of, nature – and therefore history. When in power, these dreams tend to produce the kind of collective narcolepsy from which we are waking now. The original inhabitants of our continent recorded their journey through nature in the mythic landscape of the Dreamtime. Whitefella dreaming can also walk with nature and keep open the pathways of human alternatives ('lights on hills', etc). Some seek repair, a restoration of nature, as with Martin Luther King's vision of man as a species freed from the absurdity of racism. Dreams only become dangerous when we refuse to analyse and interrogate them, and when they generate not inspiration but an intellectual slumber.

At a time of shocking revelations, it is a rich coincidence that a great Australian film, *Wake in Fright* (1971), should be rediscovered and re-released in 2009. The film brings it all together: hubris, waste, despair and natural loathing; a society that some time back took a very bad turn towards the worst possible landscape. The film retells in gothic fashion the story of a people 'still settling Australia', to echo the Australian scholar Stephen Dovers. The heartlessness and pointlessness that make *Wake in Fright* so dreadfully compelling are drawn from our wider struggle to find a heart, to find our nature, in a continent that came late to modernity. Historian Graeme Davison:

> At the threshold of the 21st century Australia has suddenly come down to earth. For two centuries our national imagination was dominated by dreams of conquering and subduing a land we always perceived as somehow alien and hostile. We wanted to explore, clear, tame, cultivate and exploit it. We wanted to mould it to the purposes we had brought with us, rather than respond to those it suggested to us itself.[3]

It may be no bad thing to wake now in anxiety if it means emerging from a dreamtime that didn't acknowledge the Australian earth and refused the hard-won wisdom of its guardians.

An underlying proposition in this book is that supernatural dreams may exercise great power to shape, even control, collective consciousness and purpose for a time, but inevitably are swept aside by nature itself. Since the Industrial Revolution we have behaved like Prometheus, the titan from Greek mythology who scorned the natural order and imagined himself greater than the gods. Neoliberalism is one such Promethean dream, casting aside nature and rapidly expanding, through globalisation, the space of resource extraction and the terrain of waste. We are living with the fearful consequences of that mad project now. Our future has been reduced by it.

The imaginings that stand the test of time are, logically, those that do not refuse history or nature – ideas like human solidarity, our dependence on nature, the possibility of failure and the frailty of human endeavour. Both left and right marched away from nature in the 20th century, succumbing equally to Prometheanism and a hubris that claimed the power to stop history. I believe it's why both imaginations now are shattering in the face of a nature that will no longer be scorned.

I've voiced concern about the state of Australia's political culture today, surely a time of greatest need for new ideas and directions. Absent vibrancy and verve, we can expect another night of the zombies, be this resurgent neoliberalism or some new ideological brake on our senses. There has been talk for years, throughout the long neoliberal night, of the need for alternative imaginings. Their time has really come. The scale and speed of the threats facing us, especially as we live in this most vulnerable of continents, make the reinstatement of our political culture urgent. We desperately need new steering to take us from the path of unimaginable pain, perhaps even destruction, that we have set ourselves on.

Lifeboat Cities is my contribution to the renewal of political culture and to the rediscovery of thoughtful purpose in Australia. The book takes the 'urban' as the starting point for any discussion of political culture in Australia. We are a nation of cities, one of the world's most urbanised peoples. This has long been so. We should embrace our long-cherished urbanity and make it central to any debate about national development.

As with any such work, the best that I can hope is that it will bring a few ideas that I regard as compelling into wider public discussion. The most important of these for me is the need to finally and utterly leave behind the Prometheanism that has betrayed our best intentions since the rise of the modern industrial ideal. At the heart of that dream is the belief that the market can replace nature as a mechanism for governing human ambition, even thought. We now, at this overripe stage of neoliberalism, even speak of a 'market of ideas'. The term perhaps best describes the contemporary university and its most apparent failing. It will win me few collegial friends to argue that the market, so conceived, has driven us to the edge of the cataclysm. There is little enthusiasm for this proposition in the 'market of ideas'.

Our world is entering an ecological emergency that now seems inevitable and which will propel us into a struggle for survival. The deep roots of the threats, and the vast changes needed to see them off, demand transition to a new social model. Not a green idyll, but something fashioned from what we have here and now. Our great suburban heartlands, which many idealists hope will vanish, will instead be the frontlines of change. More importantly, they offer a great adaptable resource from which we can fashion sustainable communities.

This book considers the principal social and ecological threats facing Australia, especially climate change and declining solidarity. It outlines some ways in which these will need to be confronted to secure our immediate survival. And yet we should already be discussing and striving to shape a social form that would bring us to permanent safety on this rich but vexing continent. We must enter our struggle for survival with an idea of something better – at least a societal model less prone to crisis than modern industrialism. Pushed back on our haunches, we will be closer to the land we have struggled to settle. My hope is that this experience will grant us time to fashion and deploy values that can guide a more resilient Australia.

The underlying theme is that we need, as a nation, to shift our collective ambition away from endless, mindless growth and towards a new dispensation favouring solidarity, care and natural renewal. This means waking from the Promethean dream in which we imagined ourselves masters of the natural universe, able to plunder, command, and if necessary repair, nature at will. To deal with the threats already breaking upon us we will need to focus on achieving resilience. The storm is already here, and we must make for safe harbour, for resilient ground, as quickly as possible. For an urban nation, this means making our cities safe vessels in which to navigate the coming storm of natural and social stress.

An important and provocative theme is that the climate emergency is a crisis of over-production, not over-consumption. The book will try to make sense of this difficult idea, which underlines our inability – as yet – to wean ourselves from an economic and social model that demands constant growth in output. From this perspective, contemporary debates about our over-consuming lifestyles miss the point: it is not possible to live and consume sustainably in Australia within our present economic model. And, just as provocatively, I will argue that these consumption debates mask a more disturbing reality: that we are *under-consuming* some of the goods and services that are needed to ensure human wellbeing and, ultimately, the flourishing of nature. We under-consume care, for example. Our current systems of care are based on a radical undervaluation of the tasks of nurturing, education and development. Much of our 'care' is life support for people stressed by an over-productive and narrowly consuming society.

A new Australia is inevitable and necessary. Its location is no mystery. There is no prospect in the 'elsewhere'. We stand on the ground of this new society. It may not be here yet, but we will reach the new Australia sooner than we think, and before many would like to. The next world is unfolding; crossing to it will be painful and fraught with risk. My hope is that we can craft this world in our minds now and take steps to make it happen, even as we fight for our species' survival. This book offers my thinking on the best of new possibilities. It conceives a world that will subordinate the economy to human need and nature's balance. It will produce to sustain, not enrich. This next world will be based on the values of care, repair and renewal, not accumulation and consumption. It will shed the mantle of complacency – the idea that increasing 'wealth' will solve problems and resolve all contradictions. A new Australia will take responsibility for itself by caring about, not ignoring, the inevitable consequences of human frailty and dependence on nature. We will move from complacency to care.

The book is in three parts, which convey three progressive movements of thought and argument. They are in a time sequence: past-present; the near future; and the distant world. The first part – 'Coming through slaughter' – considers the contemporary legacy of neoliberalism and the origins of the natural crises and social stresses confronting Australia. It is a work of recording and analysis. The next part – 'Learning to see the Earth' – begins the work of political imagination. It starts by charting the pathways to safety and finds all easy exits blocked. A time of stress and rationing is inevitable. A new political dispensation is imagined, the 'guardian state', which must govern hopefully in a time of fearful necessity. It is an era of un-knowable length when the values needed for a good society will be rediscovered and refashioned, but not without pain and conflict.

Some people will see 'nanny state' when I write 'guardian state'. These are people with maternal issues I'd rather not explore. By 'guardian' I most explicitly do not mean a cosseting paternalism. I mean government committed to guardianship of the common good, of common resources and of collective hope in a time of extreme stress, when there will be every temptation for many of us to split from the herd in a pointless bid for individual salvation. The splitters will perish, in one form or another, and splitting will weaken us when we need to be strong. We also need a state to guard against the various forms of criminal selfishness that can thrive in 'war-like' times. But none of this must extinguish the hard-won political freedoms that have marked the modern journey. We will have to surrender consumption liberties and be dissuaded, forcibly if necessary, from collectively harmful activity.

But the freedom to elect and shape governments and for cultural and individual expression must be maintained, even enhanced, in our quest for social resilience in hard times. Let's not delude ourselves that contemporary market societies represent the highest forms of freedom. This book commits us to a higher view.

The book concludes by essaying the next world, to which we are now travelling. This is a work of political imagination and hope. My view is not catastrophist. I believe firmly in the prospects for this new Australian society. It may puzzle or surprise some that I return to the project of modernisation as the basis for this battered but better world. As part of this I recover and reassess the work of Erich Fromm (1900–80), a European thinker known to earlier generations of psychologists and philosophers. He wrote in portentous times, under the 'smoking volcano', as he put it, of Nazism rising in Germany in the 1930s. Fromm, now unfashionable and unheard, buried under layers of postmodern posturing, drew attention to our tendency to fail or twist the modern ideals that framed the Enlightenment. Nazism was a spectacular wrong turn into barbarism. But Fromm believed that consumer capitalism was also wrong, a grim cul-de-sac off the road to human freedom. The current penchant for climate (and other) scepticism must also be judged as a betrayal, not an affirmation, of the modern quest for human enlightenment and freedom.

This final part of the book is a new subscription to the ideal of modernity and its glittering prize, human fulfilment. I restate my belief in the human endeavour that sought to free us from the grubby toil and cloudy thinking of pre-modern society. The fire of crisis will burn away the new illusions we took up on this great journey. The braided magic of the market will flare away, as sparks fly to heaven. A brighter, harsher world might even be good for us. Without the burden of dreams, we mortals might make the best of it.

COMING THROUGH SLAUGHTER

We've discovered the knack, the shaman's trick:
how to get right inside
of life, just when life's not looking,
to screech hymns backwards in four-poster beds,
to fool the Gods
and spit in their sober Homeric faces; to hoodwink
the Grim Reaper and Santa Claus, to April-Fool
the cheap journalism of Time …

David McComb, 'Beautiful Era' (in *Beautiful Waste*, 2009)

EVER SINCE READING MICHAEL ONDAATJE'S BOOK *Coming Through Slaughter* (1976), I've waited for a chance to use its title in my own work. Ondaatje's story, which traces the life of jazz pioneer Buddy Bolden, explores the modern predilection for 'creative destruction'. It was the Austrian economist Joseph Schumpeter (1883–1950) who gave us the latter term, to describe the paths that markets take to innovation and growth, paths that are inordinately expensive in terms of the dislocation and wastage they leave in their wake. 'Creative destruction' has a certain arty playfulness to it, but Schumpeter used it to summarise a terrible reality which goes to the core of the modern capitalist experience. The growth compulsion is a beast that must be fed with live bodies and precious resources. As part of this, we have learned to discount the awful human cost of change, especially politically forced change that has sought to clear the way for growth. Countless lives have been tossed into the furnace of reform and innovation. No one really ever stopped to ask them whether it was worth it.

Reflecting on the past few decades – the years of reform and then of the miracle economy – I've found my chance to deploy Ondaatje. We were told it was a time of wondrous growth. We were also reassured that the 'fundamentals' were right and that we could look forward to perpetual growth. The system failed on both counts. First, even during the boom years, growth was sluggish and uneven. Then came the great default of 2008/09, which showed this wondrous machine to be a Ponzi scheme of economic trickery and political manipulation. At what cost? Many lives were sidelined and discounted. This was nowhere more evident than in our cities, where new netherworlds of poverty have formed.

It was an era of fantastic waste and destruction. A charnel house of growth, again reflected in our eco-cidal cities. Our long-running despoliation of nature was speeded up and we now face the legacy of both industrialisation and our more recent experiments with neoliberalised capitalism. New gales threaten, and this time they are not under our control and not in the service of the growth imperative. Climate change and resource insecurity are the greatest of these natural tempests. We emerge from the consumption carnival into the face of the coming storm, weaker not stronger, denuded of resources and of ideas. The twin ecological and social defaults that have opened so dramatically in recent years implicate two institutions as pivots of system failure: markets and governments. All this shatters the dream of liberal democracy, the simple faith that weakly steered markets could guide us through history.

We've come through a slaughter of riches and possibilities, to a new realm of human vulnerability. This is no catastrophe, yet. The times compel us to rethink and reset our states and economies in quest of sustainability and resilience. We must pass through this era of adolescent self-harm and move into a new maturity, where we can live peaceably with nature and with our own roiling ambitions for freedom and realisation. The dictum of human adulthood might be that 'There is no creativity in destruction.' No more 'shaman's tricks'. No more spitting in the face of idols. Have we learned this yet?

2

'FIRE IN THE HEAVENS'

Fire in the heavens, and fire along the hills,
and fire made solid in the flinty stone,
thick-mass'd or scatter'd pebble, fire that fills
the breathless hour that lives in fire alone.

Christopher Brennan, 'Fire in the heavens' (in Poems (1913), 1914)

THE WORLD IS IN THE GRIP of a frightful economic recession. The suddenness and the scale of collapse resemble those of the Great Depression. Australia has survived the turbid seas of global recession relatively well. We are not, however, unscathed; unemployment and underemployment have ballooned, many have lost savings and assets, and a large proportion of households teeter on the brink of mortgage default as interest rates begin to rise again from 'emergency' levels. Our population is booming in the midst of the crisis, swelled partly by emigrants from recession-plagued countries, including the United Kingdom, Ireland and New Zealand. Our long-neglected infrastructure is struggling to cope with the human growth surge. A sense of vulnerability shimmers around the relief that Australia has ridden the recession well so far ...

'Where did this come from?' many ask in bewilderment. It seems only moments ago that we were all celebrating a boom that seemed unstoppable. And why wouldn't it last for ever? Economists told us that we had 'the fundamentals' right after decades of difficult, often painful, economic reform. The price of new wealth was insecurity and overwork, but many of us thought it worth it. Property wealth rose steadily, there seemed to be work for everyone who wanted it, homes were stacked with new toys, even miners were paid like kings.

During the late 1990s, a new name appeared for a class of go-getters: the 'aspirationals'. It was nothing more than a new term for the working-class improvers. Unlike the ungrateful, restive poor, who occasionally rioted and usually refused to behave, the aspirationals revelled in the new prosperity. To be trapped in poverty, in public housing, in unemployment, apparently meant that you had no 'aspirations'. The aspirationals moved in waves from stodgy blue-collar suburbs in the big cities out to the McMansion frontier, where behaviour, as well as buildings, was masterplanned. Both sides of politics cheered them on and joined in the sneering at 'urban elites', whoever they were.

The scripted aspirational order emphasised independence, materialism and freedom from the 'nanny state'. It was a fictional script that merrily ignored the state subsidies flowing to the new 'affordable' private schools that flourished on the urban fringes and the vast amounts of easy credit and mortgage debt that kept the whole show moving. The early years of the new millennium saw the suburban consumption carnival reach its brassiest pitch. The show would surely go on forever ...

Then suddenly it all went very, very bad. In 2008 the Global

Financial Crisis swept through the world like an avenging horde of locusts, stripping all before it. A grinding economic slowdown followed, throughout 2009. The worst predictions have not come to pass, or not yet, but the economic and social losses have been great. In late 2009, a large proportion of Australian households were perilously close to the financial edge. There were then 230,000 more unemployed people than there had been a year ago. Many have been pitched from full-time to casual work, a fact not apparent in 'improving' economic figures. The divide between the underworked and the overworked is widening again. The squeeze will only get tougher as prices for basics – electricity, petrol, water and food – continue to rise significantly; that much is certain.

In June 2009, author and commentator John Birmingham led us into the netherworld of the newly unemployed. His essay took us through the shattering experiences of three people, 'Jack, Havock and Dusty', each of whom had been made redundant in the wake of the crisis. These three unconnected individuals were drawn from different class backgrounds: Jack was a graphic designer, Havock was a facility maintenance worker, and Dusty a former executive PA whose husband had been recently laid off. Their diversity underlined the pervasive ness of employment insecurity in the neoliberal age. Before this great recession, insecurity was already the theme of our working age. Hard work could not protect you from the caprice of a flexible, impulsive economy. Birmingham writes of the cost, 'In lost wealth, and broken lives. Havock, Jack and Dusty do not deserve this. They were none of them slackers.'[1]

Meanwhile, the worrying spectre of climate change began to make its presence felt. Years of prolonged drought produced a series of urban immolations, beginning with the Canberra fires of 2003, which put sudden heat on the sceptics and the indifferent. The terrible Black

Friday fires of February 2009 came perilously close to Melbourne's suburban edge. These cataclysms – economic, social and natural – seemed to rise up from the very ground we'd considered safe and prosperous to remind us of what insecurity really meant. No, the fundamentals weren't right, it seemed. On the contrary, we seemed *fundamentally wrong* about our presumed omnipotence.

But was this all really so shocking and unpredictable? Beneath the hubris of the economic boom there were always warning signs that what we were doing was not sustainable. A few brave commentators questioned the sustainability of debt-fuelled growth. Others opposed the idea that all was well in the aspirational heartlands of our boom cities and regions, pointing to rising homelessness, the widening wealth divide, the epidemic of overwork and the declining wellbeing of our children.

Fiona Stanley AO relentlessly drew our attention to the fact that our kids were getting 'fatter, sicker and sadder', while adult assets (and waistlines) ballooned. The problem was noted in other Western countries: the simultaneous growth in affluence and decline in children's health and happiness. Commentator Anne Manne lamented our new worship of work, and the devaluing of care and human dependency that seemed to accompany it. Were kids, our most precious dependants, simply 'collateral' in the battle for material success? Something was fundamentally wrong with the growth model, but you would never have known it through most of the 1990s and in the early years of the new millennium. The mainstream press and its commentariat derided these critics as doomsayers.

In many quarters, it was heresy to speak thus. A hallmark of the times was to quickly and roughly tag any critic of social conditions as 'negative', 'miserable', or just 'not focused on the positives'. These are labels that many social scientists were ordered to wear during the

Long March of the 'reform' years. During John Howard's rule, it was especially effective in cowing the universities, and the entire endeavour of social science. Scholars reporting the increasingly obvious trend towards social polarisation were described as 'negative hand wringers', or were charged with political conspiracy. Climate scientists copped much of the same during Howard's aggressive reign of scepticism. David Marr details the persecution of public intellectuals by the Howard government; my Griffith colleague David Peetz, a pro-worker advocate, was labelled a 'union fruitcake'.[2] However, as Marr points out, the culture of clamping down on expert dissent pre-dates the Howard era; earlier Liberal and Labor governments have both sought to intimidate critics and dissidents.

My book *Australian Heartlands* (2006) aroused the happy police. It argued that all was not well in suburban Australia, and I predicted that the aspirationals would be left high and dry as the 'miracle economy' receded and the debt dunes swept over them. A reviewer in The *Courier-Mail* dismissed the book as 'gloomy and dysfunctional' and suggested that I 'get out a bit more'. The weekend supplement in the same issue ran a cover feature on homelessness. It was emblazoned 'Welcome to the Smart State, where 26,000 people are homeless; where any low-to-middle income family is three, maybe four, bad turns from the gutter.' It was a stunning and disturbing story from a gifted young reporter, Trent Dalton, recalling the Victorian slum journalism of London's Henry Mayhew and Melbourne's incomparable John Stanley James, pseudonym 'The Vagabond'. I read Dalton's account of his journey amongst the urban homeless with mixed feelings: exhilarated by its insights into our vulnerability to personal collapse, disheartened by the palpable sense that this was Victorian London, Melbourne, whatever ... on replay.

The contemporary media didn't know or care much about the

army of lost souls haunting our cities and suburbs. In 2003, *The Australian*'s Christine Jackman wrote about the monstrous rape and murder of 5-year-old Chloe Hoson, who lived in a caravan park in Sydney's west. I've often cited this rare instance where the raw underbelly of the boom was exposed, in all its vicious indifference to failure and suffering. Jackman took us behind the crime to the life-world that the little girl had inhabited all too briefly. She spoke to other residents of the caravan park, strugglers consigned to life in what they said was an utter 'shit-hole'. Their home, in effect a welfare camp, was in one of the poverty sinkholes that have opened in the struggling parts of our cities and regions: it's an area littered with syringes, lingering prostitutes, junk food outlets, servos for fags and expensive milk, but no fruit and veg shop … Jackman gave the park residents a voice to fume with. 'Dave' spoke with impotent anger about the impossibility of keeping kids safe in such a den of drugs, pollution and bad behaviour: 'Beneath this father's fury is a deeper, brooding resentment at the powerlessness of life on the fringes of Australia's wealthiest city.'[3]

While John Howard was prattling on about the aspiration and 'choice' he was delivering to his very own battlers, the *real* battlers were having fate dished up to them cold. 'There's nothing here, mate,' 'Dave' told Jackman. The young man who murdered Chloe, Timothy Kosowicz, was a mentally ill cannabis user who had been through the revolving door of contemporary psychiatric 'care' many times. When he was found not guilty by reason of mental illness, the judge declared that the community had failed Chloe: policy failure had been heaped on policy failure, with murderous result.

We continue to live with the terrible legacy of our failing care systems. In 2009, the *Courier Mail* exposed yet another urban encampment of lost souls, including many people with mental illnesses who were lacking basic health and housing support. The Alpha Accommodation Centre in Aspley, a northside suburb of Brisbane, was a violent repository of the neglected and the angry. A police union spokesperson described the park as 'out of control' and a threat to local businesses.[4] This sad portrait of lawlessness and abandonment describes many such caravan parks across metropolitan and regional Australia. Many, like Alpha, are targeted for redevelopment and closure. No, not a humane solution to the residents' woes, just the property market at work. But who is paying attention? We've been mesmerised by the marvellous ever-escalating property market of the boom years.

And yet beneath the hubris, beneath the heavy layers of indifference in mainstream public culture, there was a sense all through the boom that all was not quite well. We knew that there was a price to be paid for the new wealth, and that it wasn't borne equally in the community. Many people in precarious employment were working long, grinding hours at cost to themselves and those around them. There were the poverty sinkholes and regional struggle towns that gave the lie to the claim of universal affluence. More homeless people were lingering in our urban netherworlds. Most counts put their number at around 100,000. The rising tide was clearly not lifting all boats. These things were not invisible. We knew about them. Many of us were witness to them in our increasingly stressed daily lives.

We weren't alone in having these doubts. Robert Reich, in his influential book *Supercapitalism* (2007), saw Americans in 'two minds'. On the one hand, they wanted the lower prices and the rising asset values that came from a deregulated, profit-driven economy. The interests of shareholders were the interests of everyone, the mantra ran. On the other, however, Americans were troubled by the underside of this diabolical bargain: the corporate sector's exploitation of workers and the environment, and its general the rejection of social responsibility. This included a rapidly declining sense of obligation to employees – hours could be slashed, work shifted to inconvenient hours or offshore, in the drive to lower labour costs and maximise shareholder return.

This sense of doubt resonated in other developed nations, including Australia, as the 'supercapitalism' model spread its influence. We supercapitalists were part of a new world order based on the principles of neoliberalism, a political philosophy that prioritises profit over basic human values such as solidarity and care. Its advocates aren't necessarily sociopaths (at least in intention). They just have a cheerfully simplistic view of the world: maintain economic growth at all costs, and the rest (society/nature) will look after itself. Kevin Rudd recently read the death notice for neoliberalism, declaring 'The time has come, off the back of the current crisis, to proclaim that the great neoliberal experiment of the past 30 years has failed, that the emperor has no clothes ...' I'm not sure that the Emperor is ready to leave the stage.

Then there was the mounting evidence that nature wasn't happy with us, to put it mildly. Even through the 'miracle' years from the mid 1990s, growing numbers of Australians were increasingly disturbed by two comets that seemed to be streaking across and spoiling the bright skies of prosperity: climate change and oil scarcity. One fiery trail reported a climate cooked and despoiled by human greed. The other marked the disappearing trail of a vital resource, the energy that propelled us to greatness, and yet ultimately became our downfall. Both entwined menacingly above us, one glowering with rising strength, the other fading and falling away.

The heavens aroused and inflamed are an awful force. Their anger began to shake the groundwork of everyday life – our jobs, our holidays, our hobbies.

The very earth upon which we stood seemed to be moving under our feet; things, solid things, seemed to be swaying around us. Our wonderful climate – the envy of the world – seemed to be turning on us. *Terra Australis* was morphing into Terror Australis, a blast furnace of drought, heat and capricious tempests.

All this before the financial and economic shocks that rudely interrupted the growth carnival in 2009. The warning signs of natural stress were there: the salinity crisis, collapsing biodiversity, coastal erosion, wilder bushfire seasons. There was the bone-grinding drought that brought much of the southeastern coast of Australia to the brink of environmental default. Southeast Queensland went right to the edge; closer than many of its own inhabitants, let alone the rest of the nation, realised at the time. Politicians looked genuinely scared as they announced ever tougher restrictions on water use.

At a time when affluence was venerated, the nation was gripped by concern about scarcity. Not scarcity of good domestic help, Chilean wine or smart European ovens. It was water, the fundamental means of existence, that we appeared to be running out of.

In April 2007, then Prime Minister John Howard intoned gravely that the nation's food bowl, the Murray-Darling Basin, might soon fail. There was talk of the need to import food. By June 2009 there was relief: minor rains had granted the Basin's human dependants another 12 months of water supplies. It's a pity about everything else that needs water. Desperate times are now becoming the norm, it seems.

Even in cities, traditionally immune to drought, years of prolonged water shortage showed in the greying, lifeless gardens of suburbia, where there lurked a quiet, deepening gloom about the deaths of things once cherished and nurtured. The rains returned to many parts of the continent in early 2008, but the big dry continued in Victoria and parts of New South Wales and Queensland. This manifestation of climate change told us, should we not have already known, that urban water crisis is now a permanent spectre in our cities.

Meanwhile, oil, the lifeblood of our economy and everyday lives, also turned into a problem. It became harder, more expensive, to keep a grip on lifestyles based on cheap petrol and unrestrained mobility. The petrol shock eased with the onset of the economic crisis, which sent oil prices southwards in 2008. But pump prices still remained at historically high levels and more people were finding it harder to pay for basics, at any cost, as unemployment and cutbacks spread.

For the last few years, an odd, rather unlovely new term, 'peak oil', has crept steadily into public conversation. It is, we are told, the looming moment when the world's oil reserves will start to decline. The idea has been about for a while, but has been dismissed by governments and industry as the baseless rantings of survivalists, doomsayers and eccentric dons. Not so any more.

Both the Australian Senate and the United States auditor-general have accepted that the peak is real and imminent.[5] No matter when it occurs, global demand and geopolitical instability mean that the golden age of oil abundance is behind us. Chevron admits that 'the age of easy oil is over'. In January 2008, Shell chief executive Jeroen van der Veer predicted a global fuel crisis in just seven years.

Since the first fire in a cave, access to energy has defined human existence. We learned to be pretty careful conservers of the stocks we had. But modernity put an end to that quaint practice: more fossil fuels could always be found, and technology could transform them.

Peak oil is shattering the perpetual motion dream of the carbon economy. For most of us, the oil default is sudden, unexpected and deeply inconvenient. The busy free-ranging lives celebrated – indeed mandated – by neoliberalism are threatened.

Aspiration is turning to desperation. In early 2007, a survey of more than 5000 Australian families identified rising petrol prices as the main source of financial concern.

Sometimes passing through and surviving one (modest) crisis engenders not a sharpened wariness but its opposite, a heightened sense of indestructibility. So it seems with the 1970s oil shocks, which by the 1990s had passed comfortably into memory. The theory that market societies were the 'end of history', our highest and most invulnerable social form, survived.

So the unexpected return of oil scarcity feels deeply unsettling; it cracks open a cemented faith in our invincibility. All the more unnerving is the mounting evidence that coal, our other great – if these days unseen – energy source, is fuelling climate change. Most of us are guiltily aware that Australia is a global 'filthy man', stoking the carbon economy with cheap, dirty coal. It's dashed inconvenient that exporting it doesn't distance us from the problem, or ultimately from blame.

In the latter years of the miracle boom these shocks and shifts increasingly disturbed the political climate of Australia. They surely helped turn the electoral tide against John Howard, the boom's most cheerful advocate, in 2007. His stubborn eco-scepticism looked more and more out of time and place as the unpredented drought ached on year after year and as the evidence of global warming became more compelling and more accepted in public discussion. Then there was our 'other mind', to use Reich's terminology, which turned against laissez-faire approaches. What was the point of consumer affluence when WorkChoices threatened us with more insecurity and fewer breaks?

Even before the big financial bust of 2008, there was evidence of a deep unease in Australia's social consciousness, which is the substrate of politics. Electoral shifts and opinion surveys showed that many of us weren't buying the argument that all was well. A new social sensibility became evident. There was rising awareness of our exposure to sudden, even wild, changes in the basic forces that industrial capitalism had considered vanquished, pacified and shackled to the wheel of progress.

This is where the 'two minds' problem kicks in for most of us. It is why the headache that began in the good times is becoming a migraine in the slump.

We are intimately aware of, and buoyed along by, the economic wave that carried most (but not all) of us to material prosperity: the jobs, the toys and the travel opportunities. When the times were good we knew at the personal level how to manipulate our own role in the economy for personal gain. But most of us had no immediate connection to the big forces at play in environmental change, and thus little sense of how to comprehend and intervene in these processes. Global warming is … well, global. What can we do about it? Increasingly the answer is that we have to live greener lives and consume less. A mountain of books and guides now cheerfully tell us how to live green lives, but many miss the point: it's the economic system, not our lifestyles, that is the root of the problem.

The green living prescription seems to have two glaring weaknesses. How can we be sure that *everyone else* is consuming less and consuming responsibly? And is cutting consumption a good thing when the economy is nosediving and the government is urging us via stimulus payouts to spend more? Shouldn't we consume to save jobs? A new and more painful form of the 'two minds' dilemma confronts us: no longer is it a trade-off between affluence and ecology or fairness. The choice now seems to be between our immediate livelihood and our long-term sustainability *as a species*.

This perhaps explains the muted, confused atmosphere of politics in 2009. Governments are using their full force to secure the economic basics, but acting with great hesitancy on the fundamental threats of climate change and resource insecurity. Science is telling us that these threats will, to use the language of Hollywood, very soon be 'clear and present' dangers. Perhaps the terrible Victorian fires of 2009 already tell us that the beast of warming stalks our land. If so, we will have to acknowledge its presence, not its imminent arrival. And that means reframing the terms of survival and sustainability. Jobs and consumer goods won't be much use in a brown and blasted land.

We're not alone with this dilemma. A global economic crisis is colliding with a global environmental crisis. I think the West is experiencing a moment of rude rousing from many dreams. In Australia, it's as if the whole dormitory has woken simultaneously, angry nature shattering the windows with sudden force. What happened to the warm narcosis of the miracle economy? How did the dream of freedom morph into a schlockbuster about global warming and oil vulnerability? Who brought this horror upon us? As consensus grows on the threat posed by global warming, the tendency to blame and punish also grows, as Howard found out. For some time, a swathe of the urban commentariat in Australia has been blaming the people who punished Howard's inaction on climate change. They have a seductively neat answer to the question of culpability: suburbia. Suburbia is the consumptive beast whose appetite has ruined us all.

This is an idea I will challenge later because it is misleadingly simplistic and it urges us towards some very inequitable responses to climate change and resource insecurity. Just keep in mind for now that the wealthy, who bear most environmental and economic responsibility, do not tend to live in the suburbs lambasted by green critique. Many are living the outwardly green lives in inner-city apartment land. But this green presumption is a ruse, and I will tackle it too later.

It's 2010, and the economic sky has fallen on our heads. In a recession-plagued world, Australia has survived relatively well, but with deep injuries, many of them hidden from view, in the poorer netherworlds of our cities. The ground beneath our feet is shakier than ever. And our natural world appears to be edging towards a dangerous tipping point. None of this was supposed to happen. The miracle economy unleashed by the Hawke-Keating reforms should have swept all obstacles before it, if only through its sheer power to throw money at problems. But of course this never happened. Much of the 'wealth' we celebrated now seems fictitious, wiped from balance sheets in the blink of an eye. And the investments in the future never happened. It was all left to the market and we can see the results of that malign neglect now, in the social injuries and the environmental depletion that have left the Earth with truly terrible and terrifying dilemmas.

What will happen to the billions of earthlings trapped in impoverished, poisoned Third World nations? And what of our own ruined communities and the many more now teetering on the edge? Looking back, the boom looks much less innocent. Indeed, the words 'boom' and 'growth' seem misplaced – too fertile and jovial. 'Emerging from the boom' seems ludicrously lighthearted. What of the human and environmental resources misused? The social alternatives laughed at and dismissed? Are these resources and plans still waiting in the wings now that we have a crisis, or have they been expended and extinguished?

The fire in the heavens cannot be extinguished by economic recovery. Prosperity may or may not be restored for a time, for some. It's no longer the main issue. We cannot restore an injured biosphere or renew depleted resources without first experiencing a time of great trial, threat and reassessment. We have arrived at the edge of a great default, and are now walking backwards into a very uncertain future.

3

CITIES ON THE EDGE

THE GLOBAL FINANCIAL CRISIS (GFC) put a mighty spanner in the neoliberal works. The global growth machine jammed up in 2008 and looked as if it would fall apart completely. Since then Australia has, as noted, fared comparatively well compared with other countries. Our little cogs have continued to whirr within a very sick machine. Yet our own pain cannot be underestimated. Australia has registered significant job losses and a growth in underemployment. In September 2009, doom merchants Dun & Bradstreet reported that a third of Australian localities (by postcode) had fallen into 'high risk' of financial distress, meaning they were likely to be hotspots of mortgage default. This landscape of vulnerability had grown by 30 per cent from the previous year.[1]

There is tentative talk of a 'recovery' in late 2009, but near consensus that more pain lies ahead. Unemployment is expected to rise in the context of a 'recovery' overshadowed by rising interest rates and rising costs. What is not acknowledged, mild upturn or not, is the flagging ability of Australian households to bear new waves of volatility. As I write this, in November 2009, my local paper announces in a blazing front page story that 'Almost 60,000 people in southeast Queensland cannot afford enough food to eat each week.' It tells me that 'tough economic times [have created] a new class of "working poor" struggling to feed their families'.[2] The focus of this and many other local urban papers is shifting from roads and rubbish to the increasingly undeniable problem of working poverty. It's very possible that a 'recovery' will simply lock many weakened and vulnerable households into employed penury, as is the case for many in the United States.

Decades of restructuring and lopsided growth have left a significant sector of the population without the energy or the resources to endure further stress and uncertainty. There's a powerful Salvation Army ad doing the media rounds which draws attention to the social injuries accumulated before the GFC. It depicts a sad young girl on a news board carrying the headlines: 'ECONOMIC CRISIS. It's all you've been reading about. It's all she's ever known.' GFC or not, we have legions of metropolitan and regional communities that have been hammered by decades of restructuring and polarisation. The chronic volatility of a market-driven economy will soon be overlaid by the certainty of climate change and its impacts on everyday life. Add to this the looming resources shortages – of water, food and oil especially – and we have a picture of cities and communities whose resilience may not stand much further testing.

The continuing pain of the downturn is unlikely to be evenly distributed across the population or across our communities. But surely the legacy of the miracle boom is that most communities have been fattened up enough to lose a few kilos to a recession? This is a sadly misleading view that ignores how vulnerable much of Australian society is to the current downturn. For many communities, recession is not a threat; it's all they have known for a long time.

George Megalogenis, in his influential book *The Longest Decade* (2006), posits the idea that the recent economic surge bucked long historical trends by outlasting the usual 10-year cycle of boom, then bust, then boom, etc.[3] It's a brilliant book that makes many insightful observations about the neoliberal project in Australia. But *The Longest Decade* tends to overplay the boom and understate the extent to which it failed to reach many Australian communities. Many people and places went backwards, not forwards, during the growth carnival. How will they fare in newly tough times?

Perhaps the gated wealthy enclaves that now pepper major cities such as Sydney and Brisbane will weather the storms of financial and economic crisis.

But what about the communities that didn't fare so well during the past decade? Some are hardly small – much of Adelaide shared a sense of prolonged recession and hardship during those years. For many regional centres, especially those dependent on industry, such as Newcastle, Whyalla, Geelong and Wollongong, the miracle economy is a parallel universe they haven't visited yet. And our larger cities contain many poverty sinkholes that are as much a testament to the unevenness of the growth surge as are the bright new master-planned estates of affluence.

Clare Martin, CEO of the Australian Council of Social Services, recently painted this picture of lingering urban poverty:

> As we breathe a sigh of relief at signs that the Australian economy appears to have dodged a bullet, we cannot forget the 2.2 million Australians who are still in the firing line. That's the number of Australians who are faced with the daily, inescapable reality of poverty. That's one in every 10 people, including 412,000 children, who live below the poverty line. Being poor means going without. Skipping a meal to pay for your child's school books, leaving the electricity bill unpaid or

asking for yet another extension to pay the rent. It means
constant worrying about how the week's budget can be
stretched, or having to borrow money from friends or payday
lenders to replace the car or fridge.[4]

Analysing deprivation in the eight capital cities using 2006 census
data, Griffith University's Scott Baum exposes what he terms the
'suburban scars' of our decades of economic restructuring. More
than 10 per cent of localities were highly deprived 'poverty traps'
where multiple disadvantages – impoverished neighbours, poor
services, dispiriting surroundings, isolation – have combined to
lock inhabitants into long-term poverty. At the other end of the
deprivation scale, Baum finds a wealthy mirror image: just over 10
per cent of suburbs contain high concentrations of wealth, either
in established older areas of affluence or in the newer bastions of
privilege in the inner cities. Within suburbia, the same polarities are
revealed. Gated republics celebrating order, similarity and privilege
are juxtaposed with neighbourhoods studded with syringe-littered
parks, payday lenders, sex shops, takeaway food outlets and two-dollar
bargain shops.

These metropolitan sinkholes include some of our most neglected
public housing estates, where many socially dependent people have
been crowded together and now languish in the twilight of anxiety
and disrepair. Public housing in Australia used to serve a much broader
segment of the populace, including working families. In the late 1970s
a well-intentioned but disastrous decision was made to target only the
poorest and most disturbed members of our community.

Since then, people with the highest social needs have been herded
together in crumbling public housing estates which had been designed
for working families. Positive role models were few, and the design and
layout of these estates included many areas where people could not be

seen by others. Kids roamed without supervision, parks became increasingly run down, a minority of 'bad neighbours' terrified the rest and whole communities began to spiral downwards in a cycle of fear and neglect. Some public housing estates avoided this fate, but many didn't.

Brave and sometimes successful attempts were made to renew estates; in southwestern Sydney especially, some marvellous successes were recorded. Claymore, a crime-ridden estate, was turned around by a shift to community management. A community garden became a focus of pride and industry and a local employment scheme found work for tenants. But these small courageous initiatives battled a larger tide of official neglect and indifference. During the Howard government's tenure the Commonwealth showed little interest in the cities or in the public housing sector. States bumbled along with declining resources and flagging interest.

Meanwhile, the social outfall from more than 20 years of economic restructuring began to mount in our cities. The neediest, or sometimes simply the luckiest, managed to get a hold of a public housing place. Lucky? The alternative was to join the swelling ranks of the homeless roaming our cities. Many of these are children and young people. About half are under the age of 24. Some 70 to 80 per cent have mental health issues.

The Rudd government's unprecedented determination to tackle homelessness is admirable, but the task of dealing with decades of national social neglect is truly daunting. And made more so by the economic crash. The waves of people becoming homeless now are becoming so as a result of the recession, not of family breakdown, mental illness and abuse. Their surging numbers in 2009 caused NSW state MP Paul Gibson to urge the opening of army camps in western Sydney to house the homeless.

Those able to avoid the streets waited on ever-lengthening public housing waiting lists. Meanwhile, they usually ended up in crappy private rental housing. The worst of it included the urban caravan parks I mentioned in the previous chapter. But outside these gated poor communities, other whole neighbourhoods were sinking into the economic sands. Think of the desolate congregations of 1960s and 1970s 'six pack' unit blocks in the working-class regions of our cities – north and western Melbourne, western Sydney, etc. Again we were herding the poor and the needy together, but this time in private housing, not public housing.

These areas were characterised by old, poor-quality rental housing, much of it overpriced, a crumbling public domain, and a lack of public transport. They tended also to be 'food deserts': places where wholesome produce was hard to find but junk food was in abundance. Many households had children, and it's hard to imagine a worse place to rear kids. However, the urban poverty holes that were formerly caravan parks come straight to mind. The one I mentioned in the previous chapter, the Alpha Accommodation Centre in Brisbane, houses about 200 people, most of whom have been referred from the prison or mental health systems. A social worker quoted in the *Courier Mail* observed recently, 'I find it heartbreaking that children live there and are surrounded by a community that is abusive, neglectful and provides few roles models and little hope.'5

But all this was largely ignored during the boom. Those who pointed out the growing divide between the urban rich and poor, and its harmful consequences, were branded doomsayers: 'A rising tide of growth will lift all boats', 'Beware of the politics of envy, don't criticise the rich', etc. As I pointed out in the previous chapter, my own 2006 book, *Australian Heartlands*, attracted this sort of reaction from many critics.

During the early years of the new millennium I lived and worked in western Sydney with my family and experienced first hand many of the wrong turns we were making in terms of urban and social development. We spent a year in a newish but poorly-built flat just north of Parramatta, the ever-battling, neglected twin of Sydney. Parramatta was a hub of human services, which had to care for more than their fair share of Sydney's downhearted and downcast.

Our flat was an artefact of the market-driven urban consolidation that had racked and ruined many parts of Sydney during the previous decade. There was almost no sound insulation: 'You can hear your neighbour change her mind', a taxi driver once told me. We sometimes heard worse. There were many other design faults that made family life a challenge: leaking water pipes, no safe play space for kids, etc. During this time I endured the ghost trains of western Sydney, travelling daily to my workplace in the southwest at Macarthur. For me it was another disturbing reminder of the perverse fact that the economic boom had been accompanied by a gross deterioration in Australia's urban public realm, especially in the poorer parts of our cities.

It was 2001. Transport for the Olympic Games worked wonderfully; there was good organisation, and it was helped by lots of residents leaving town or driving less. In the months after the flame and the tourists left town, I felt, with increasing despair, that the well-oiled machinery of the Games had never been intended for normal domestic service.

It was all coming back: the countless delayed or cancelled trains. The worst times were the winter return journeys, when the darkness added chilled austerity to the whole tedious, confusing experience. As the time to leave work approached, the dread would build as I thought about what lay ahead: windswept East German-style platforms; the garbled who-the-fuck knows announcements; the interminable waiting for a train that should have come; skulking hoods in the dark; the fearful station staff holed up in their prefab bunkers; the wafting smell of piss and vomit; the few sad others, usually migrants, braving this mindless trauma.

Fellow travellers avoided eye contact, though sometimes I'd catch a fleeting stare. Meanwhile my thought balloon said, 'Will I get home tonight? Will I see my son before he goes to sleep?' How many other families were stressed and separated by dysfunctional transport systems – and this includes road congestion – and by the lack of affordable, quality housing? A large section of the middle class shared the pain of Sydney's growth: rocketing housing prices, congestion, cramming, pollution and a generally overcooked life.

Eventually it was too much and we did what most other flat dwellers with kids were waiting to do. We fled for the outer suburbs and traded the leaky cubby house for a home in a new estate in Sydney's southwest, near my workplace and right in the heart of aspirational country. It was an interesting transition, from a place where no one spoke to you much, to a place where neighbours took a keen, occasionally intrusive, interest in what you did, how you kept yourself.

We had the anonymous note in the letterbox experience at one point, for some infraction of the aspirational order. I think I'd failed to mow the lawn regularly enough. There was a vaguely present sense of unspoken rules being assumed and policed. I related this to a newspaper journalist interviewing me after the publication of *Australian Heartlands*. Incredulity was the response: not here in Australia, surely? Doesn't that *Truman Show* stuff only happen in America? She sent a photographer to the estate we now lived in (against my wishes), to get some images for the story. A few days later I received an email from the journalist informing me that when the photographer got out of his car and took photos, local residents rang the police. The journalist wondered whether I'd had a point.

All this is to point to the mess we got ourselves into during the so-called boom. Too many people had been left behind and, worse, were crowded into poverty hotspots where disadvantage and despair were compounded by proximity. At the upper, or at least aspirational, end of the scale, we were herding families into *Truman Show* estates, where homogeneity and closely observed 'good behaviour' was the norm. How could either of these community forms be good for kids, or indeed anyone?

Now the gales of recession are upon us. The poor have little defence against them and many in the debt-geared middle class are highly vulnerable. By May 2009, almost a quarter of Australian mortgage holders were experiencing 'financial stress'. That's about 1.3 million *households* and a fair chunk of the population. 'Financial stress': what a pathetically administrative way of describing the countless scenes of daily family horror that bedevil our suburbs, the rows, the glowering despair, the cutbacks, the blame, the cowering kids … and for many, the messy, messy end of it all.

First-home buyers are really copping it – 30 per cent are suffering 'mortgage stress'. By early 2010 it was obvious that many who had been lured into mortgages by the federal first-home owner grant were close to the line or had crossed it and fallen into default. It's probable that the high transport and energy costs of running a home in a far-flung new release estate had pushed many over the edge. Access (or not) to public transport can now, as never before, decide a suburban family's fortunes. Perhaps for some of them, there aren't yet kids involved. Or, God help us (and them), they are too young to understand the dissolution of their family lives. The recession is shaving a few points off the Gross Domestic Product (GDP) and some fat cats are having economic haircuts. But the frontline is made up of the mort-gage battlers, now in debt and – often – in housing that is bigger and more expensive than they need, in public transport-poor places, usu-ally on the outer fringes of our cities.

'Oil vulnerability' deeply threatens the suburbs, which, unlike inner metropolitan areas, have limited access to quality public transport services. The analysis of metropolitan oil vulnerability carried out by my Griffith colleagues Jago Dodson and Neil Sipe shows how car dependency plus mortgage debt plus low household income are together rendering the burdens on outer suburban communities in Australian cities unbearable. As their book *Shocking the Suburbs* (2007) demonstrated, this dire threat was even then being manifested in Australia's cities, and especially in its suburbs, as global oil prices surged to historic levels.

The onset of recession in 2008 added serious injury to the insult of 'petrol pain', especially for the many self-employed contractors who are the life supports of working and lower middle class suburbia. These people – tradies, domestic helpers, home deliverers and the like – are utterly car dependent in ways that many of us don't comprehend. The car isn't simply a means of getting to work, it is largely their *place* of work. The cheerful assumption that fuel would be endlessly cheap is no laughing matter for them as they struggle with the increasing costs of car ownership and use. The soaring greenhouse emissions of our cities will eventually have to be reined in, perhaps savagely, as I will suggest later in this book, because we seem destined to leave our response to the last possible moment. This will open a new energy divide in our community, with great potential to harm suburbanites who are dependent on non-renewables, especially petrol, to maintain their everyday life.

Amongst the ranks of the vulnerable, it is surely children and young people who are most exposed to the injuries of recession. In 2009, some 50 per cent of the people helped by southeast Queensland food charities were children.[6] The situation is surely repeated in other cities and regions. Evidence suggests that family breakdown is rising as unemployment and austerity mount.[7] This is one of the worst 'injuries' that kids can suffer. Their resilience in stressful times should be our first priority.

One concern is that rising mortgage foreclosure and unemployment will increase crowding in our cities. The United States is well down the track of recession. The Australian commentator Guy Rundle reports that in the United States 'multiple occupancy – moving in with your parents or your kids or sharing, or sleeping in your cousin's living room – has risen by 45% in the last 3 years'. In April 2009, some 324,000 US households received a foreclosure notice. About 1 in 5 American homeowners was in negative equity, meaning that their mortgage was larger than the value of their house. The negative equity epidemic was expected to get much worse. Much of middle America exists in a twilight world of scraping by and smouldering frustration. The poor remain biblically always there, a grimly permanent fixture. This didn't happen overnight, but reflects the long and divisive influence of neoliberal thinking on that country. In addition, the wealth grab by elites in the United States over the past few decades has been protected by a deep cultural aversion to progressive thinking. One result of this is that the poor are usually the first to blame themselves for their situation. The GFC has made worse a problem of social polarisation that has been building for a long time.

It's doubtful that Australia will experience these levels of mortgage default and crowding, because our home lending system wasn't given over to market chaos to the same extent. But there's plenty of evidence that housing pain will be long and deeply felt in Australia. Again, not

43

an encouraging prospect. This suggests the need for an aggressive affordable housing program, especially in already stressed cities such as Sydney, Melbourne and Brisbane. It means weaning ourselves from a tendency to see housing as wealth for some rather than a right of all. We must find new ways to house a younger generation that, sadly, seems resigned to the idea that it will never have secure accommodation.

Our cities are on the edge of environmental default. This is exemplified by the desperate recourse to desalination for drinking water in Perth, Adelaide, Melbourne, Sydney and Brisbane. And at the same time, social tolerance is straining to breaking point in some urban communities. Policy neglect and faith (in the freewheeling market) brought us here.

But some of the criticisms of this dangerous urban state are wrongheaded. As I will show in Part 2, the suburbs are *not* the overwhelming source of environmental overload, as much green critique supposes. But they are most prone to the consequences of our looming crises, especially oil default and the pain that climate change will inflict. The wealth of the inner cities (and richer suburbs) renders them safer – for now – from the storms to come. It also, however, implicates them deeply in the origins of our deepening problems. The consumption lifestyles of the rich have large environmental footprints. Justice will demand that they bear the largest share of adjustment 'costs' and pain.

So much for the 'urban blame game'. It's a serious issue, because the apportioning of fault and responsibility has been skewiff until now, and has generally let the rich off the hook. But ultimately we're *all* on the hook. Getting to the deeper sources of the twin environmental and economic crises will drive us to re-examine and rethink the very foundations of our social system. Our consumption patterns are a manifestation of, but not the root cause of, the problems we face.

Our biggest problem is *overproduction*. We don't know how to produce less. This is the unspoken secret of our modern existence. All the green wash and green goodwill aside, we are a society apparently hardwired to massive and ever-expanding production. It's all very well urging greener consumption, but try arguing that we should slow or reverse production. It's the fastest way possible to the political 'Exit' sign. You can't stop the slaughter. We face a crisis of overproduction but no one wants to acknowledge it, let alone stop it.

4

A CRISIS OF UNDERCONSUMPTION

… the built environment is bleak. The East Derwent highway
slices the sprawling estates of purpose-built, grey-block houses
into four awkward sectors dotted with lonely Metro bus stops,
splotched with vandal-proof paint jobs. Bridgewater's shabby
shopping centre is marooned on top of a hill. Which heavy
young mums and older pensioners struggle to reach on foot.
There is a bargain store called Chickenfeed, a supermarket
and a charity shop, its windows covered with metal grilles.
In Gagebrook, homeboys on bikes cluster around the graffiti-
smeared front of a fish and chip shop. Bigger blokes hot rod
around suburban cul-de-sacs, or fold their arms with attitude
as I drive past front yards cluttered with car wrecks. I suspect
they're up to no good. Druggies and dropouts for sure …[1]

Natasha Cica, *Griffith Review*, 2007

THIS IS AUTHOR NATASHA CICA'S SNAPSHOT of two poorer
communities in Hobart in 2007, well before the onset of the GFC.
Fast forward to now, 2010. There is much handwringing about
the problem of our overconsumption in the lead-up to the global

economic crisis. We put everything on the credit card and whooped it up, and the downturn is our comeuppance for years of waste and frivolity. Add to this the gross consumption that apparently lies at the heart of the global warming crisis. Green critique waved a censorious finger that was ignored during the consumption carnival. Moralists shook their heads at the cult of materialism, but the writhing masses were indifferent. These criticisms enjoy a new authority in a suddenly chastened era. Universal thrift, saving and circumspection seem the necessary medicine.

But how does this square with Cica's portrait of urban exclusion and need, written at the height of the growth carnival? It could have been written about many such communities in any of our cities. We hear lots today about overconsumption – and about consumers behaving badly. I think we need to cut through this to recognise that two crises of *production* intersect in the contemporary global failure: of social *re*production under neoliberalism, and of *over*production under longer-run carbon capitalism. The overproduction argument will be dealt with in the next chapter. Here I wish to consider the alarming underconsumption of socially nutritious things during neoliberalism. Natasha Cica's grey snapshots from the boomtime album bring back to mind the great social fasting that has long plagued many urban communities.

Why were these urban famines so resolutely ignored during the boom? The odd urban flare-up, such as the Macquarie Fields riots in Sydney in 2005 briefly broke the spell of complacency. Part of the explanation is that there was a general political reassertion of an old 'blame the poor' game: it's their own fault for not getting on in good times. The aspirational debates played into this by casting the poor as uninspired laggards.

But in my view there was something else that was specific to the neoliberal era, and that suppressed awareness of dysfunction in the system: our cities allowed the withering of democracy as they were turned into growth machines, economic engines in a global competitive game. A series of booms – in construction, IT, housing, population – tended to drown out the bad news, especially the failing will and fortunes of many poorer communities.

Try to recall any systematic and open debate about where our cities were headed and in whose interests. There was precious little. It was an era steered by business elites and neoliberal state governments, all wedded to a model that geographers have termed 'urban entrepreneurialism'. Cities were cast as corporations fighting each other for economic spoils on the national and global stages, and requiring a firm hand and good business sense. In spite of the corporate rhetoric, much of the focus was on the playgrounds and embankments of the elites in CBDs and inner cities, with the suburbs largely cut loose to weather the tempests of change on their own.

We'd do well to ponder where such urban processes are taking us in a time of economic and environmental threat. A newly urban species, *homo urbanis*, will mostly encounter these threats in cities and large human settlements. Australia and its cities don't yet feel the searing pain that is so apparent in global pivot points such as New York or aspiring tigers such as Dublin, where unemployment soars as public finances plummet. But as I argued in the previous chapter, our cities are already on the edge of social and environmental defaults that widened during the era of neoliberal rule.

The tempests of global warming and resource (especially oil) volatility threaten the fundamental circulation and health of urban development that yearned for ever higher velocities. This process wanted to free us from our oldest foe, time. Time is the agent of delay, decay and debility, which remind us that we exist in nature, not outside it. Neoliberal reformers seemed to think the city could be freed from nature through the *annihilation of time by space* – in this case, the space of the city. This is to reverse the formulation of Karl Marx, who, a century and a half ago, observed that capitalism was a volatile expansive force that deployed technological advance (railways, machines, etc) to overcome all barriers, including topography and distance. In Marx's era, through the expansion and improvement of steam shipping, railways and telegraphic communication, technology was used to accelerate commercial time and overcome spatial barriers, reducing the tyranny of economic distance.

This process has continued since, and with increasing impact on the whole human species. Consider economic globalisation and the air travel and telecommunications that have driven it forward. The internet, like nothing before it, is 'instant time' that annihilates spatial frictions and links communities and people in a great web of simultaneity. Of course geography still matters – as geographers like to say (we would) – because even the internet cannot overcome political geography. Consider the filtering that many countries, such as China (and now possibly Australia) apply to the web.

Nested within this global crunching of space by time, we see, at the urban level during neoliberalism, the reverse – the use of urban space to speed governance, the economy and everyday life. Cities are spaces of control, where governments and powerful corporations often act together to exert authority over change. In our system that means state governments, corporations and powerful industry lobbies. Think of the road-building coalitions who have worked with the states through murky devices such as Public Private Partnerships to force through unpopular changes to our cities: tollways and tunnels have torn up urban fabric and cost enormous sums of money.

Sydney's tunnel debacles in the last few years only partly managed to expose a 'tollway industrial complex' that has exerted enormous undemocratic power over our cities during the neoliberal phase. Several of these wasteful, unneeded schemes have gone broke; most recently (in January 2010) the Lane Cove tunnel. Brisbane is presently embroiled in a tunnel and tollway construction scheme of epic proportions. I think this TransApex scheme will prove a very costly and unpopular act of folly. Decades after we have become aware of the danger and damage of car dependency, the production of roadspace in Australian cities continues unstopped, and seemingly unstoppable.

The metropolitan space has been governed in the interests of, and by, those wanting to speed things up. Such interests wish to keep decision making – about infrastructure, about planning, about basic urban priorities – ahead of public comprehension, beyond the ordinary politics of democracy, which neoliberals tend to despise as a brake on the things that they believe matter: growth and profit, narrowly conceived. Neoliberals see democracy as a hoarding behind which lurk deliberation and dissent, those infuriating societal tendencies that delay the supremely significant work of production and gratification.

Economic velocity is vital to neoliberalism – we need 'chronic activity' to keep us busy, we need the system to keep on furiously humming in a way that forestalls any awareness and discussion of the mounting balance sheet of social and environmental debt. This was achieved in an urban space that ignored time: cities. Cities were fortresses of speed, of youth, of debt deferred. By contrast, in many rural and regional areas time could not be vanquished and indeed became more salient, as depopulation, decay and decline set in.

Recall the slogans and language of the neoliberal urban project: boundless mobility, frictionless exchange, easy credit, 'can do' governance, policy 'going forward', big build infrastructure. Its most memorable emblem was the Kennett government's (1992–99) slogan

'Victoria on the Move', which was emblazoned on everything for a time, including number plates and the signs at the exits to Tullamarine Airport. Jeff's claim was misleading: Melbourne may have been 'on the Move', but increasingly disaffected country folk saw only rural services slowing and disappearing.

Back in the city, though, it was all go: tollways and ring roads were added to the urban landscape, and a new city cathedral, the Crown Casino, was built. *On the Move*: this was not time for consultation and contemplation of alternatives. The 'urban', in short, was a space where time was refused. This was a famine for democracy from which we haven't fully emerged. Our cities are hardly humming with democratic discussion. On the question of citizenship, the portly aspirational consumer looks positively malnourished, even skeletal.

This was an order that spoke much about time but did not *value* it. And yet ultimately time caught up with it. The GFC is a temporal crisis, a crisis of flighty, fictitious wealth at last brought to ground by the 'real economy'. The infinite resource time of neoliberal urbanism now looks unexpectedly finite. A stop clock has been set above the oil pump. Water is more strictly allocated. By one means or another, energy is asserting its 'real price'.

The new wisdom is that financial meltdown, and the economic recession which emerged from its wings, are crises of indebtedness: of overconsumption and under-saving. The environmental crises are even more straightforwardly the spawn of excess consumption. And yet these assessments are missing something fundamental. Claims of overconsumption miss the origins of economic and environmental malaise by overlooking the *under*consumption of necessities, especially social values, under neoliberalism.

Let me define 'social values': they are the fundamental resources and activities that ensure that society is able to function smoothly and to 'reproduce' itself. That is, the goods, services, expressions and values that nourish humans, human relationships and social wellbeing. They are largely things that the market doesn't value, unless it is bribed to do so by public subsidies. Included are care of those in need, and the constant, demanding task of nurturing our young. Add to this the arts and cultural expression, which inspire and enrich in ways that mainstream economic activity never can. These are the values, the activities and the commitments that mean we aren't a historical dead end; that Australian society is able to continue with an acceptable degree of harmony, wellness and happiness.

I've already mentioned our malnourished democracy, which is surely a threat to our future. Beyond this, we also threaten our future – and just as fundamentally – by ignoring the imperative of care. A society is a 'social corpus', a body that must attend to its developmental needs, its inevitable wounds and infirmities, and to the constant yearning of its heart and soul. It must make accommodation for its weaker parts and tendencies in order to keep the whole strong – the dodgy eyes that need glasses, the shorter leg, the weaker ear, the failing memory, the skittish temper. And these supported parts aren't just passengers; they too provide sense and sensibility to the whole.

Care is *the* great unheralded social enterprise. All of us will be amongst the ranks of carers or the cared for at some point in our lives. Sometimes we will be in both groups at the same time. Consider the elderly wife, caring for her husband with dementia in their home, both receiving support, such as meals on wheels, and you get a sense of the vastness and complexity of the picture that we will all at some point be involved in.

This is the social part of us that was pushed off stage during the carnival of neoliberalism. And this heedless brushing aside isn't a trivial act; it's a death wish, because social reproduction isn't an optional exercise, or something you can put the pause button on. If you upset the long-running process of generational change and handover, you upset the whole social apple cart. And there is cultural evidence of this 'cart tipping' in the form of lots of angry commentary about the 'privileges and arrogances' of the baby boomer generation. This was colourfully signalled by Ryan Heath's 2006 book *Please just f*off: It's our turn now – holding baby boomers to account*. Increasingly, as those following have found voice, they evoke not the usual generational criticism of the elders as conservatives, but as vandals and thieves who made little provision, *who cared little*, about those who were to follow.

The first fault line of neglect is the widening disparity of wealth under neoliberalism, both in Australia and internationally. As British analyst David Harvey points out, 'since the 1970s the policies of neoliberalism have been about wage repression'.[2] He sees this as part a wholesale process of 'accumulation by dispossession' that has been central to the neoliberal project. In Australia, labour's wage share of national income has decreased by over 12 per cent since 1975, while the share to profits has increased by nearly 50 per cent. This shift has been reinforced by changes that directly or indirectly effect wealth transfers, such as the privatisation of public assets and deregulation of working life. It has been reflected in a new urban geography of disadvantage and segregation. Its lost constituencies include the new army of the homeless, and the over 700,000 who are on the Disability Support Pension.

New ruptures in the social fabric of reproduction include the poverty and crime hotspots I discussed earlier. There are rising social and health morbidities in poorer communities. Don't discount the vastness of the last problem. Just one dimension of this is the rampant growth of Type 2 diabetes, which is nearing epidemic proportions in some parts of the country. Health professionals refer to Type 2 as 'lifestyle diabetes': a preventable condition that thrives in communities where diet, stress levels and exercise are problematic. In the last five years in Queensland, for example, the number of diabetes 'hot zones' (places with abnormally high incidence) has risen from 1 to 69. There are 200,000 people with Type 2 diabetes in Queensland, and many more who don't yet know that they have the disease. It has claimed the lives of 3000 in the last five years.

The steadily advancing recession is revealing a more generalised legacy of social vulnerability. As I discussed in the previous chapter, mortgage failure is spreading amongst the suburban middle class, especially those in the 'aspirational' echelons who were persuaded that they could manage high levels of household debt. A failure of planning, and of public fiscal provision, has also left these middle and outer suburban regions without public transport. Increasing numbers of people in these areas live a great distance from where they work, but both public policy and the employer mindset have ignored the dire consequences of these ever-lengthening commutes.

Planning that was obsessed with issues such as density and design tended to ignore the vital issue of jobs–housing balances in communities – that is, ensuring people live close to where they work by dispersing affordable housing stock and directing employment to accessible hubs within metro regions. The deeply embedded cultural presumption of 'boundless mobility' has played a major role in this planning failure. The celebrated suburban improver, the 'independent contractor' – is now feeling the crushing burden of (commuting) time.

Ostensibly, the neoliberal project aimed to rescue the individual from long slumber in the twilight of the welfare state. And yet it did everything but that. An age of radical individualisation that opposed individual dependency ended up reproducing it massively at the social scale. The awful truth for neoliberals is that after the long march of reform, nearly 20 per cent of us are receiving income support from the government. Before we set out, about three decades ago, it was much less than 10 per cent of us. And now many of us are financially reliant upon this assistance. In its pursuit of structural economic reform, neoliberalism in fact generated a rate of social dependency that has made the state more crucial, not less crucial, to everyday life.

In Australia before the great slide of 2008, the average consumer could afford much of the product coming from countries where wages were low and environmental costs were ignored. But many critical mainstays of health and wellbeing were unaffordable. In the era of 'Made in China' affluence, you could snap up a DVD player but you couldn't get your teeth fixed. The evidence was seen in indicators of fundamental wellbeing, such as oral health, in the general populace, and especially amongst poorer Australians. *The Australian* newspaper recently reported, 'More than 650,000 people are on public dental waiting lists across the nation … with the dental health of children worsening and rates of tooth decay and gum disease well above many other developed countries …' The 'Lucky Country' lost its smile somewhere along the way.[3]

The Brisbane writer Scott Pape observes:

> The consumption boom that made us all feel rich was driven largely by Americans using their homes like ATMs. They took the inflated equity values from their homes and used the proceeds to buy stuff from China.[4]

We did much the same.

In the wake of the global crisis, much has been made of the overproduction and overconsumption of 'fictive economic value'. This means goods and assets that have fictional worth, not fixed and defendable value, such as housing. Much of the consumer goods that were brought within reach of the middle and working classes during this era carried fictive *human* value: they were largely unrelated to basic human needs. The expanding flow of household consumer goods did little to aid the basic cause of social reproduction.

Fictive value was also manifest in the collective consumption areas recast by neoliberal reform. In particular, human services, including health, disability, childhood and aged care, were remade into forms that entrenched social vulnerability. In the mental health and disability fields, state dependants were reconstituted as 'consumers'. Many Western countries did the same from the 1970s as they deinstitutionalised and restructured human support services, adopting a 'provider-consumer' model that replaced state 'care'.

This sentiment was justified, but the strategy was wrong. Older, mouldering institutions and services had to go – many were brutish and an affront to notions of decency and human rights. But recourse to market models rather than communal and not-for-profit models effectively threw many of our vulnerable babies out with the old welfarist bathwater. In areas such as child care and aged care, which are publicly subsidised, the rise of profit-seeking service providers led to a volatile and heartless landscape of 'care'. In mental health and disability there was a double jeopardy – consumerist models without the public subsidies to make community care a reality. The vivid illustration of this failure is the legion of mentally ill living on the streets of our cities and towns these days.

Ideologues were certain that marketised human services would restore the individuality and dignity of the cared for, who would enjoy and deploy all the self-regulating powers of the consumer. Social provision and support were expanded, to be sure, but it was all undermined by market delivery. As the failures of these 'reformed' human services were revealed in Australia and internationally, this was shown to be a false dream. The limits of this neoliberal approach were never more apparent than in fields such as early childhood and aged support, where publicly subsidised corporates produced no better than miserable, and ultimately crisis-prone, modes of care. Despite driving costs – and conditions – downwards, the corporates could not extract sufficient profit from care, even with generous state subsidy.

The 2008 collapse of the Australian multinational child care provider ABC Learning epitomised the failure of social reproduction under neoliberalism. In late 2009, a Senate inquiry into child care damned the profit-driven approach of Australian child care, which it held responsible for the ABC catastrophe and for poor standards generally. We created, it seems, a child care 'industry' that does what markets do: pursue ceaseless growth with small regard for human consequences. The fact that mindless growth too often ends in collapse – social or environmental – was graphically demonstrated in the convulsive expansion and bust of ABC Learning.

The developmental needs of children have been discounted by the lust for economic gratification. The evidence presented by leading child health expert Fiona Stanley and others shows the magnitude of the threat: the 'present generation of children may be the first in the history of the world to have lower life expectancy than their parents'.[5] The simultaneous escalation of national wealth and childhood morbidity has been observed and remarked upon in many developed countries. In North America this perverse trend has been termed 'modernity's paradox' and the 'American paradox'.

Neoliberalism resulted in a gross *under*consumption of social values. By condemning consumption generally, the 'progressive' politics emerging in the wake of neoliberalism neglects both the looming crisis of social reproduction and the misery already endured by many groups, notably the homeless, the mentally ill, the unemployed and the swelling ranks of casualised workers.

I haven't addressed many other areas of care that were subject to malign neglect during the growth surge. These include parental leave, foster parenting, drug and alcohol rehabilitation, young people's development and support for people subjected to sexual and other forms of violence.

The absence of care nurtures its deepest foe, despair. I was struck last year by a story in the *International Herald Tribune* newspaper. It concerned the assisted suicide of a Bettina Schardt, a retired X-ray technician from a city in southern Germany. She ended her life with the assistance of a German Dr Death, Roger Kusch. Bettina was 79, but not ill or disabled. She was alone and lonely and dreading the eventual prospect of a nursing home. Kusch put an end to her despair, and because suicide is not illegal in Germany, he will doubtless help staunch a rising tide of elderly despair in the same way. The article that reported Bettina's death carried a picture of her, in her last days, eyes blazing with dark, despairing intensity; an image that will remain with me until my own demise, I think. Eugen Brysch, director of the German Hospice Foundation, stated in the same article: 'The fear of nursing homes among elderly Germans is far greater than the fear of terrorism or the fear of losing your job.'

Who can doubt that many older Australians share this dread? Is there a link between the regular exposés of horrifying failures in aged care and the growing advocacy for euthanasia? Even during the GFC, there are fates worse than death. And what of the miserable worlds we have created for many of our children? An investment in future human desolation? Even before the miracle economy began to falter, at the height of the fantastic boom, despair was lurking and growing. During the boom we unlearned many things we previously and quietly knew, including how to care. Only a blinkered analysis can view this as a time of overconsumption. Perhaps the best assessment of the boom is that we had fun while it lasted but at the same time brought a plague on our own houses. As any party-goer knows, overindulgence usually goes hand in hand with a failure to care for ourselves. In these hungover times, we look back at a binge that nourished few and wasted much.

5

A CRISIS OF OVERPRODUCTION

IN SEPTEMBER 2009, TWO GERMAN SHIPS became the first commercial vessels to navigate the Arctic's normally frozen North-East Passage.[1] The shrinking ice cap confirmed many scientists' worst fears about global warming. A study released the next month predicted that ice-free summers would be the norm in years to come. Most worryingly, Arctic warming is a threat, not merely a sign. It has the potential to unlock and unleash the vast amounts of greenhouse gases stored in the permafrost soils. These are said to contain double the amount of CO_2 that is in the atmosphere.

This scientific alarm was countered by new commercial opportunity: a new Arctic freeway would greatly speed the journey of ships between Europe and Asia, shaving 10 vital days off voyage times. The bright impulse of globalisation will not, it seems, be blunted by environmental collapse. Captains of ships and captains of industry will haul us through this new passage of opportunity. But is it reassuring to know that the lunatics are still in charge of the asylum? And that now the asylum is burning ...

I've argued that consumption is the wrong starting point for any consideration of climate change. There are two reasons for this. First, and most pragmatically, all recent attempts to moderate and make less harmful our consumption patterns appear to have amounted to nothing. Resource depletion and environmental despoliation continue unchecked. As a species, we are now consuming considerably more of our natural capital than we were 20 years ago, and this after decades of green critique and the embrace of sustainable lifestyles and designs. There are plenty of instances where resources and places have been conserved, in the face of an ever-ravenous consumer culture, but at the global scale, our economy is literally costing the Earth. The second problem with a consumer-led approach is that we remain implacably committed to economic growth – or at least our leaders do – and so the relentless exhaustion of our natural resource base continues unabated.

This latter problem goes deeper than the boosterist sentiments of our leaders, who evoke sustainability on the one hand and boundless economic growth on the other. But it's worth examining the depth of this duplicity. A burning symbolic instance was the billboards erected in 2006 by an avowedly green Queensland government at Brisbane Airport. In the midst of widening community panic about a grinding regional drought surely driven by climate change, these placards depicted great clambering cranes making their way across a dusty landscape. It boasted, 'Head to Queensland, the Climate's Great for Growth'. But was growth good for the climate? As social observer Peter Spearritt asked:

> So in Queensland, the climate is good for holidaymakers, for locals and 'for growth', the optimum trilogy. What could go wrong? No-one ever suggested at the countless focus groups run by political parties that the cities might run out of potable water.[2]

The truth is that the Queensland state government, like all other liberal democratic governments, was fulfilling its most basic and unyielding obligation: to maintain a market economy that is hardwired for growth, not conservation. Even the tag 'liberal democracy' is rather misleading and perhaps ought be discarded in contexts where economic relations are prioritised over everything else. As with other Western nations, we commonly describe our society as a 'market economy', saving other ideals, such as 'social solidarity' or 'commonwealth', for textbooks and official letterheads. In the era of neoliberalism, the reference to these other civilised social ideals was progressively stripped away, leaving only what has, after all, been the driving force behind social relationships for nearly two centuries in the West: the market.

We can see this most basic urge and duty straining away in government responses to the twin global threats of warming and recession. The Australian national government yearns for an emissions trading scheme but also thrusts itself decisively into the task of economic stimulus – boosting consumption. Through 2009 a stark difference in levels of will was evident. The timid, studied embrace of weak carbon reduction strategies contrasted with the steely resolve that characterised efforts to restore, or at least shore up, growth rates. The abiding anxiety is that for every carbon reduction measure there is a 'jobs trade-off'. The conservative Opposition shouts out that formulation, and the Labor government dreads it. Environmental modernisers urge that it isn't true and that a greened economy will guarantee employment, even boost it, but as long as we have the market as our economic bedrock, they are probably wrong.

It's an inconvenient truth that deep environmental interventions in the economy must confront the market's urge to expand. This is a confrontation that will 'end in tears' either way – through either a retraction of environmental policy ambition or the loss of jobs. In the latter case, perhaps exemplified in the programs to reduce logging of old-growth forests, the employment losses and community pain will wane over time as new arrangements and outlooks take their place. But there's no escaping the consequences and the costs of standing in the way of the market's drive for ceaseless expansion. And when we stymie this expansive urge in one economic area, it's inevitably offset by new growth in another. We may protect our forests at the expense of the forests in developing nations. Or the (unlikely) conservation of both may drive the development of resource-hungry alternative building products.

Governments in liberal democracies have fairly narrow concrete obligations – essentially to protect the market, to guarantee the rule of law and, if there's time and energy left over, to safeguard civil rights. It's hard, therefore, to conclude that their anxiety around carbon reduction or any profound environmental measure is misplaced. Criticising the obsession of states with growth and market expansion is in this sense disingenuous. They are only doing their job. And we punish them when they don't do it well. Unemployment and economic stagnation surely drive electoral behaviour far more strongly than do the spectres of global warming and resource depletion. However, environmental collapse now has a permanent place in the imagination of the citizenry. In the past we regretted pollution of places, but now we face the degradation of the entire globe. In this we, the public, are as conflicted as the governments we elect and later tear down. We fear environmental decline, but have no way of translating this anxiety to a political system that is dedicated to preserving the market as the fundamental organiser of economic life.

And we have only a limited inclination or power to turn our concern with climate change into concerted behavioural change.

Opinion polls consistently find rising public concern about warming, and high levels of support for macro counter-measures, such as emissions trading. And yet, as Mark Latham correctly points out:

> These are interesting [poll] figures, especially when compared with the number of people who, in practice, have altered their lifestyles and lowered their living standards in the fight against greenhouse gases. The statistics ... show that no more than 10 per cent of Australians have tried to reduce their carbon footprints.[3]

I don't think this is evidence of mass negligence, but it does indicate that it's pretty much impossible to live sustainably – that is, within safe emissions limits – in a contemporary Western nation. In a market system that celebrates choice, we have in fact curiously little discretion over our consumption patterns. Zero (or even low) carbon consumption is too expensive and too impractical for most of us who have to survive within an economic system that is hardwired for growth, and that propels us into ever faster, ever more thoughtless lifestyles. Think of the speeding up and the intensification of life since the advent of mobile phones.

The business sector is not so conflicted; certainly not in its corporate mainstreams. Here there is a less vexed understanding of the need to pursue the cause of growth, which starts and finishes at the company's balance sheet and share value. Robert Reich, a key figure in the US Clinton Administration, analysed the corporate imperative in his book *Supercapitalism* (2008).[4] He explains the rise of neoliberalism globally, but especially in the developed English-speaking world, as a process that lifted the covers off the underlying duty of firms to maximise profits and shareholder return in market societies.

Prior to neoliberalism, in what Reich terms the 'Not so Golden Age' from post World War II to the 1970s, the corporate sector was restrained by social expectations and obligations – to maintain high wages, workforce stability, good labour relations, and a sense of civic purpose and contribution. This was reflected in a long period of high employment levels and low rates of social polarisation. There were exceptions, and bad behaviour, but not generally amongst the industrial behemoths – they saw their vast fortunes and prospects as dependent upon a larger compact of civic improvement. The Cold War helped to focus corporate minds on the need to not drive workers into the embrace of socialism through bad behaviour. CEOs led by example, drawing modest – compared with now – salaries and self-consciously involving their companies in national development and the cause of social harmony. It was hardly an egalitarian paradise, but a restraining hand was placed on the baser instincts of the market and the corporate sector. The differences between rich and poor were muted (though real), especially in Australia.

A shining Australian example was Gus Dusseldorp, the Dutch postwar immigrant who established the Lend Lease group of companies. His firms became synonymous with the ideal of corporate civic-mindedness, and Dusseldorp lent his considerable intellect and energy to important causes, including reforms to urban development processes which aimed to improve fairness and reduce corruption. Dusseldorp served on a national inquiry for the Commonwealth that recommended, amongst other things, that development rights for outer suburban land be reserved for the community. This was seen as a means to reduce speculation and private windfalls taken at community expense – the development game that in 1979 Australian urban analyst Leonie Sandercock memorably termed the 'land racket'. It's hard to imagine a leading urban developer today agreeing to, let alone recommending, this kind of intervention.

The stagflation crisis of the mid 1970s ushered in the wolf. The sheep's cloak of civic responsibility was discarded in the name of restoring profitability and growth. This began what Reich terms the era of 'supercapitalism', where the corporate focus shifted from worker relations to consumer and owner relations and a new dedication to cost and price minimisation and profit maximisation. The ideology that cloaked supercapitalism, neoliberalism, also discarded the long-held view amongst elites and governments that social disparities were a bad thing, and ultimately harmful for the corporate sector itself. This view was tipped on its head. Now the rise of a conspicuously wealthy class would help to inspire the poor and energise the lazy. Walt Disney couldn't have spun a better fable.

Many of the older industrial conglomerates were broken up, chipped away at or even lost in the transition to a new economy. In the West the production of things also gave way to the consumption of things. Production, but not ownership, shifted to the developing world. Our firms made things offshore more cheaply and found ways to make us consume more of their product. After a period of flagging energy, market capitalism got back up off the canvas and, with more vigour than ever, began a new round of expansion. A kind of cheap prosperity ensued: what I termed 'Made in China Affluence' in the previous chapter.

However, as Reich points out, through this new era of affluence we felt an increasingly painful bipolarity. As consumers we had never had it so good; but as citizens we watched with deepening dismay the erosion of corporate social responsibility and labour standards, and the increasingly rapacious plunder of nature. In 2009 clothing manufacturer Pacific Brands announced plans to cease production in Australia, meaning the loss of approximately 1800 jobs. Production would go offshore in the drive to keep down labour costs and thus prices for consumers back in Australia. The decision was heartless;

consumer and shareholder interests were prioritised over workers. The problem is that most of us are consumers *and* workers; many of us are both of those and shareholders as well.

Much reformist effort from well-meaning greens and social advocates has aimed to reinstate corporate responsibility in our business sector. We've seen a few campaigns of this nature in Australia – against Gunns in Tasmania and James Hardie nationally – but the main action has been in North America, where consumer boycotts and grassroots advocacy have sought to encourage a sense of commitment and conscience in high-profile firms such as Nike and Walmart. Reich's argument is that this is joyful barking up the activist tree, and has no real prospect of reforming supercapitalism, because this isn't a moral issue: it's a structural reality. The only compelling corporate responsibility is to shareholders and consumers, in that order. It's absurd to speak of good 'corporate citizens' because the corporates aren't citizens – they can't vote, apply for passports or join public hospital waiting queues. They exist for one purpose: to maximise profit through expansion, expansion, expansion ... The only meaningful forum for social change is civil society, we the people.

Governments must protect markets and make their path to higher plateaus of production and profitability easy: it's in their written orders. Firms must compete and maximise their return – collaboration and long-term investment are anathema. An army of these reformers has marched to join battle with enemies that don't exist: the evil, fire-breathing humanoids that crowd Hollywood schlock films these days. The company that pushes beyond environmental and social boundaries is not evil; it's doing its job. This is not to discount or ignore the fact that there is genuine malevolence in some of the firms that do things to harm humans and living things directly. An example is the monstrous slaughter and suffering that followed the gas leak at Union Carbide's plant in Bhopal (India) in 1984. We must distinguish genuine and explicit wickedness from the ordinary duty of firms to ignore boundaries in their endless scramble for growth. We can and should legislate, monitor and educate to prevent outright malevolence, but that won't solve the climate crisis.

Given the speed and the imminence of the warming threat, and the increasing dysfunctionality of the global economy, we have to now question both the terms and ambition of governments and the sanctity of the market as a governing economic force. The ecological and economic collapse demands that we rethink and reset these two institutions, both of which have failed us. More on that in the last part of the book. Before that we have to consider how to get through the immediate crises that threaten our natural and social orders. We must pass through a valley of profound menace to our species before we can hope to win through to a more stable and humane place in the Earth's order. Much of this is the legacy of centuries of carbon capitalism – the last wild phase, neoliberalism, put the drive towards collapse on fast forward. It's understandable that there is much handwringing now about the errors of the economic rationalist turn, but taming supercapitalism won't root out the lethal paradox of our age: the underlying tendency to simultaneous social deprivation and overproduction.

The neoliberal consumption carnival that has just left town failed to meet many social needs. It was of less consequence for the environmental crises confronting us, notably global warming. The earlier industrial order that emerged in the wake of astonishingly clever technical innovations had one great flaw. It assumed access to infinitely abundant carbon.

This claim of infinite resources was resisted in some quarters. The

Reverend Thomas Malthus (1766–1834) would have none of it in his work on the 'inevitable' dangers of population growth. His ideas began a population anxiety that remains with us today. We can't discount the question of population entirely, of course, but Malthus had the wrong end of the pineapple. The real issue is how we humans organise our fundamental affairs, especially our requirement for sustenance and shelter, and meet our social needs. For centuries we have relied on the market to govern our relationship with nature. The evidence suggests that we might need a different relationship counsellor.

The Malthusians – wrongly, in my view – finger population as the root cause of the stress and likely collapse of nature. This fear was largely swept aside during the rise of industrialism by a powerful mainstream view that *technical improvement* would always push back the frontiers of nature. Technological innovation and ever higher efficiency could render a finite resource boundless. And if it finally ran out, the endless age of discovery would yield new resources and new fields for extraction.

This Promethean view was premised on the belief that nature was a force to be tamed and shackled to the wheel of progress. Industrial power showed it could be so. There were opponents who saw the rising volcano of the market, not the growth in the size of the human family, as the trigger for natural disorder and depletion. In 1883, Frederick Engels warned: 'Let us not, however, flatter ourselves overmuch on account of our human victories over nature. For each such victory nature takes its revenge on us.' Climate change, you'd have to say, is a pretty spectacular form of revenge. The old criticism of economic growth which neoliberals declared heresy seems to have the angels on its side.

For a long time, the efficiency view has dominated. Its newest manifestation is the rising faith in green technologies. Industrial capitalism has indeed proven a great innovator, pulling greater yields from fixed inputs – but the record has been uneven, and has often masked the plunder of new resource fields as declining ones were used with more circumspection. The problem was that the drive for increased efficiency never seemed to dent the growth in resource consumption.

This is partly explained by something termed the 'Jevons Paradox', which posits that improvements in the efficiency of a resource's use tend to *increase, not decrease* the rate of consumption of that resource. It was formulated by William Stanley Jevons (1835–82) in his book *The Coal Question* (1865), which noted the simultaneous rising efficiency of coal use in England and the growth in total coal consumption. This twin effect has ever operated thus, and often to our immediate benefit. It has certainly marked an improvement in the overall welfare of human populations, at least in the West. The material power of expanding markets has been greatly intensified by rising energy efficiency. Put simply, this is the creation of much more with more. This trend looks benign only if we ignore the threat to our species' survival inherent in it over the longer term.

So let's get down to it. The threat of climate warming, already manifest, is a consequence of overproduction, not overconsumption. Consumption of inputs, and of final products, is a result of and a pre-requisite for capitalism's unstoppable compulsion to expand economic activity and value. In capitalism the market is a dynamic, self-repli-cating force, and market relations are characterised by this relentless expansion, not by equilibrium. The unplanned nature of capitalist competition means that, periodically, the output of individual firms, industries, sectors, cannot be sold. These equilibriums are accidents, temporary spaces in the long march of growth. Markets obstinately drive output beyond social need and towards the precipice of overproduction. David Harvey writes:

> Capitalists have to produce a surplus product in order to produce surplus value; this in turn must be reinvested in order to generate more surplus value. The result of continued reinvestment is the expansion of surplus production at a compound rate ... The perpetual need to find profitable terrains for capital-surplus production and absorption ... presents the capitalist with a number of barriers to continuous and trouble-free expansion.[5]

This drives unremitting territorial expansion: new territory for the *extraction* of resources (human and natural) and for the *absorption* of waste – or, in a word, *globalisation*. In the converging fields of contemporary climate science and climate debate, the question of 'absorption' comes starkly into focus. As David Harvey puts it, constant expansion of productive capacity 'puts increasing pressure on the natural environment to yield up necessary raw materials and absorb the inevitable waste'.[6] Economic globalisation, given new impetus by neoliberalism, produced new terrains for resource extraction but did not, could not, expand the atmosphere. As if in recognition, it now wishes to bury carbon emissions underground. The problem of climate change has arisen from this historical overburdening of the atmospheric terrain.

The dream of green reform is that economic growth can be decoupled from this twin territorial expansion, that industrial ecology can green production. State and civil society would then be transformed by a great 'ecological modernisation' of policy and purpose. For decades these dreams have inspired renovating projects. Some resources may be indeed used more efficiently, but this husbandry cloaks a greater abandonment: the ever-escalating consumption of the Earth. We've agonised about the Amazon for years, for example, but still watch helplessly the inexorable waning of our last great forest.

The global ecological crisis – of declining resource reserves and failing waste absorption – betrays a systemic lack of interest in reform. The *Limits to Growth* thesis, published in 1972, briefly captured the Western imagination.[7] However, the barriers of caution and conservation described in it were subsequently circumvented by the growth path of globalising capitalism. In the decades since the Western environmental movement roused itself to life, natural rapine continues, and at ever greater scales: 'deforestation in the tropics destroys an area the size of Greece every year – more than 25 million acres [10 million hectares]; more than half of the world's fisheries are over-fished or fished at their limit'.[8] The Canadian social observer Ronald Wright reports:

> Ecological markers suggest that in the early 1960s, humans were using about 70 per cent of nature's yearly output; by the early 1980s, we'd reached 100 per cent; and in 1999, we were at 125 per cent. Such numbers may be imprecise, but their trend is clear – they mark the road to bankruptcy.[9]

And what of bankruptcy in an age of abundance? Australia faces the global economic and ecological crises after 17 years of continuous growth. The insightful George Megalogenis terms our recent unprecedented boom 'The Longest Decade'.[10] This boom has not, however, bequeathed a legacy of affluent resilience. In many social areas there are no excess stocks to run down. There were many spectres at the neoliberal feast, and underconsumption of basic human values abounded. Social malnourishment is nowhere more apparent than in the interstitial spaces of the nations' cities where an army of the unhomed lives in quiet desperation. Quieter still – for now – are the many unfolding agonies in the debt-geared aspirational suburbs. Will all this be righted by the stimulus packages of 2008/09 and the structural reforms to housing, health and tax policies that must follow?

I seriously doubt it, but along the way Rudd's new social investments will at least repair some of the immediate injuries of neoliberalism. Many nations are desperately rediscovering Keynesianism as a way to jumpstart consumption. It's dangerous to consume less in the global village we must now all inhabit, like it or not. OK, it's not that simple. We know we must consume less, *but we don't know how to produce less*. Will efforts to restore the toppled world economy cost us the Earth?

Decades of green censure have done little or nothing to reset the path of consumption, which has continued to yearn for higher, more trivial peaks. We may recall the philosopher Erich Fromm's 1942 warning that the destructive contradictions of modernity would reveal themselves in this manner.[11] The great unheralded cost of *individuation*, he wrote, was alienation from Earth, kin and community. This rupture would drive an exodus towards the consolations of consumerism and other compensations for the 'terrible burden' of 'self-strength'. To stand before such a flight from desolation is perhaps to stand in the way of history. Let's hope not.

6

CHILDREN OF THE SELF-ABSORBED

A WELL-READ, WELL-ANALYSED FRIEND recommended a book to me which he thought helped explain adult anxieties of our age. This was the American psychologist Nina Brown's *Children of the Self-Absorbed: A grown-up's guide to getting over narcissistic parents* (2001).[1] It sounded intriguing and promising. Could the uncertainties, the restless fluidities and the relentless search for completion that seems to define my (middle-ageing) generation be explained by past parenting? If so, very neat. I had to read this book.

At the time (2008) I was on sabbatical leave at the National University of Ireland, resident with my family at Maynooth, a town in the outer commuting belt of Dublin. I made a special journey 'up' to Dublin to find this exotic text in one of the city's many fine bookstores. I read it compulsively on the train home and in the next days. But the more I progressed, the more I realised that Brown's message – at least for me – was not about the failings of the past. I racked my memory but couldn't recover a culture of self-absorption amongst my parents' generation – not the modest middle-class circles I was exposed to, at least. No, the more I read and thought about it, the more I realised, with creeping trepidation, that her analysis implicated *my* post-1950s generation in the epidemic of narcissism and its consequences for children today. *Our* children.

I think I just make the tail end of the baby boomer generation, by one definition at least. But we adults born from the 1960s onwards were largely raised by early baby boomers and their immediate predecessors, those born prior to and during World War II. Our parents' generation did extremely well; there was a pretty steady enhancement of working and middle-class wellbeing during this time. They experienced the long boom of continuous economic growth that stretched from the late 1940s to the early 1970s, and many did very well from it. Most, however, had smelled the whiff of privation early on, through exposure to the Depression and/or war. Many had their parents' straitened experiences stamped on them in some way. I was aware of this as a child – of what people had 'been through' – but it wasn't presented to us with pity.

I don't think 'self-absorption' describes their mindset or their rearing of children, at least generally. There was restiveness and unhappiness with roles, especially as the age of freedom dawned in the 1960s. The Commonwealth's 1975 *Family Law Act* saw an exodus of the trapped from unhappy marriages. Women regained identities beyond the anonymous confines of traditional marriage, and many other values and cultural outlooks were able to flower.

But it's fair to say, I think, that these adults weren't as focused on themselves as were those who followed. It is a little unsettling to me to see so many ageing baby boomers now absorbed in preening themselves, but this didn't tend to define their approaches to parenting.

Not so for neoliberalism's children. We are the product of two intersecting changes to social values and roles. The first was cultural pluralisation – the blowing open of narrowly stereotyped roles and values from the 1960s on in the face of feminism, gay rights, environmentalism, multiculturalism and the like. The second was the great reprioritisation of economic values that followed in the wake of the neoliberal ascendancy from the mid 1970s. The result was the underlining of the individual as the centre of social life and ambition: the process I referred to in the previous chapter as 'individuation'. From this flowed a culture increasingly devoted to the needs and the anxieties of the 'self', and expending enormous professional, material and cultural energies on the repair and development of ourselves. The 'Me' generation was proclaimed in the 1960s, but most of our parents looked on with bemusement, too busy with booming babies to join the happening. The message was really taken up by the Xs and Ys who came next.

We tend to think of this as an American influence, a tide of individualism flowing across our shores along with all sorts of other cultural detritus (shallowness, short-termism, materialism). This is partly true: the rise of a globalised culture and the new reach of American-centred media certainly gave their values and habits compelling force across the globe. It generated a lot of resentment along the way, but also a lot of mimicry and acceptance. However, this neglects the extent to which we engineered the shift to individualism in our own ideological and cultural workshops, especially as the drive to embrace home-grown economic rationalism gathered force.

The new ideology broke up the public sector, and generally devalued collective endeavour. In *Australian Heartlands* I termed this attack on civic values and institutions 'The War on *Terra Publica*'. This narrowing and reprioritisation of economic and political values was mirrored by cultural shifts that subordinated the importance of social relationships and civic values. Since the 1970s, kids and young people have been reared in an increasingly atomised culture where social relationships and values are fluid, and the only thing that is really solid is the individual: at the (beginning and the) end of the day there is me.

The rise of an increasingly anxious consumerism captures these changes. Children and young people have emerged as massive new markets. This is another dimension of the tendency to overproduction essayed in the previous chapter. Teenagers have long lists of electronic must haves – phones, music players, games – that have been incorporated into their everyday culture as necessary mainstays of networking and friendships. We are now familiar with the increasingly alarmed debates about the penetration of children's life-worlds by advertising which attempts to push new, but annually obsolete, 'necessities' into the core of family and civic life. The sexualisation of children's fashion is merely the sharpest edge of the wedge that is trying remorselessly to open childhood to the claims and dictates of consumption.

Who's doing this? Who's letting it happen? Adults, most of whom were themselves reared in a culture of rising consumption and so are also up to their necks in consumerism, in many instances competing with their kids for the same toys, must haves and thrills. In our house there are multiple phone and pod chargers, data cables, mini speakers, MP3 players, games platforms and computers, but it doesn't prevent my son and I battling for access to these 'necessities' of daily life.

More generally, society seems ever more focused on a self that isn't childlike or young. The individual at the centre of our society is an adult, and it is sometimes hard to sense children even in the dusty peripheries of social consciousness. The one exception is their new tyro-adult identity as consumers. The Australian commentator Anne Manne has written powerfully of 'The New Narcissism' in Australian (and Western) society.[2] Manne writes, 'Narcissism is not just tolerated in our day and age, it is glorified ... The hallmarks of cultural narcissism are deeply woven into the fabric of our current society.'[3]

I've thought much in recent years about the question of narcissism in contemporary urban life, especially as it is revealed in the new gated and masterplanned communities that celebrate order and similarity. These new 'gated republics' respond to a deepening desire amongst the affluent and the aspirational for communities of the 'same', where just as you wouldn't be ambushed by burglars, you also wouldn't be rudely surprised by strange or discomforting people and values. It is an uncomfortable truth that these looking-glass communities aren't necessarily safer than their ordinary middle-class equivalents. Arguably, children in such estates are exposed to the risk of *asocialisation*: that is, developing social outlooks that cannot cope with the diversity and fluidity of contemporary Australia. This is a double tragedy, as our cultural and social abundance should be seen as a source of enrichment, not vexation.

One defining quality of the new narcissism is its childlessness – the adult gaze is reflected back to itself through the looking glass of a society besotted with the here and now, not what is to come. For Manne:

> We prolong adolescence, a time of self-centredness, well into
> middle age. We are skittish about children, a project that, to be
> done well, requires investments of time and energy not in the
> self, but in another human being. We delay their arrival
> indefinitely, or look around when they do arrive for someone
> else to take responsibility for rearing them.[4]

Children by their nature demand a different perspective on time. They are fragile, sensitive and often unpredictable; their 'now' is not the solid, rational and knowable 'here' of we adults. They have much to 'teach' us about the adult delusions of invulnerability and certainty. Their developmental needs, if we recognise and value them, must question our momentary wants and impress upon us the importance of the future as a horizon of social concern. And yet contemporary Australia remains transfixed with the present, and its speeding up. Our institutions, communities, built environments, values, habits and consumption increasingly betray our neglect of children and young people.

And very dangerously for children, our work obsession is another feature of this narcissistic age. Again, from Manne:

> Narcissism in celebrities is to be expected, but the most common way in contemporary society that the upper-middle-class narcissist gazes into the lily pond is through work. Work is an uncontested value in our culture. It allows a socially legitimated form of self-centredness as we pursue our 'careers' with the piety of those answering a religious calling. Work is a proxy for the self. We reward it mightily. The workaholic corporate executive who behaves with utter ruthlessness in business and at home, who lives as if his or her family or employees have no human needs, is often admired.

We are starting to learn how much Australia's workhouse culture is harming our children and young people. In Chapter 4, I referred to the trailblazing work of Fiona Stanley and colleagues which is sounding the tocsin of alarm about declining wellbeing amongst children. We have a raft of studies revealing the hurt and harm that parental absenteeism is causing in kids. This is not an argument for reinstating the captive mother of yesteryear; it is reason to think ourselves through to a less work-obsessed culture that values and invests in the care of its young. If anything, it makes a case for releasing and committing men to the task of care, even if many are initially unwilling.

By contrast, our focus on the aged and ageing has never been greater – if largely reactive and ineffectual. We dread degeneration and death as never before and our officialdom is beside itself with the prospect of an ageing society. Our neglected and failing aged care systems betray the shallowness of this anxiety and its 'econocratic' roots. The aged are feared because they aren't good workers and taxpayers, and are a burden on the 'productive'. In earlier social orders, the elderly played critical roles, as carers, arbiters, cultural repositories, cultivators of nature, and so on. They were especially important in the lives of children. We give little thought today to how we might reinstate more contemporary forms of these roles for elderly people.

Back to Ireland, where I'd begun to think about this problem of contemporary self-absorption ... Nina Brown's book made me think in unexpected ways about the problem of narcissism beyond the gated republics in our major cities. The problem wasn't a neurosis of asylum-seeking elites feeling 'urban disorder', but a deeper social malady in which we were all potentially implicated. We the people, we the self-absorbed. Could this idea of mass narcissism be linked to my academic work on our changing urban life?

Part of my work in recent years has pointed to the neglect of children in our urban environments, places that seem mindless of children's needs and perspectives. With colleagues, I've written on this problem at some length, particularly in *Creating Child Friendly Cities*.[5] My starting point is that many developed nations, especially the English-speaking ones, have for some time ignored children in their collective thinking and in their public policies.

After reading the work of Brown and Manne, it occurred to me that the roots of the problem lay far deeper than 'official neglect' – which could be righted, I'd earlier thought, by greater awareness in our leaders and better training of our urban professionals. No, indifference to children and hostility to young people went right to the core of contemporary social relationships and the self-absorption of adults. The problem and its solution thus lay deeper than the everyday processes that govern urban life and the development of cities and communities. Hectoring and challenging officials and professionals might do some temporary good but it wouldn't expose the hardened and hidden hearts of cultural, economic and political systems.

This deep self-absorption must manifest itself in many ways, not all of them necessarily obvious in everyday life. Cities and neighbourhoods have, in a variety of ways, become less friendly, even harmful, to children and to the people who care for them. Increases in density through redevelopment have often designed out kids and carers – there are no places for play, no family services, and housing is designed for the childless. Neighbourhoods riven with roads endanger children and reduce the possibilities for healthy activity. A public liability crisis reinforced by blunt health and safety strictures has dumbed down kids' play environments. Bodies corporate ban kids' games. Green spaces where wild, nourishing play occurred have been lost to urban development.

Chris Guilding, my colleague at Griffith University, has spent some years studying the landscape of bodies corporate, the unelected groups that rule the world of flats and units. He sums up the whole sorry urban consolidation story for children in chilling terms: 'It's often said in the industry that the big challenge for body corporate communities is the "crap" – children, renters, animals and parking.'[6]

I have a lecture I've given on this topic in recent years which also points out some of the ways in which we might make our cities better for kids. While in Ireland in 2008 I travelled to Yeats' beloved Sligo Town to deliver said seminar, at the invitation of the Institute of Technology. In the hour or so before my presentation I wandered the environs around the IT and chanced upon a famine graveyard. This encounter moved me deeply and caused me to ponder anew the inexplicable tendency of humans to harm their young, sometimes unknowingly, sometimes, sadly, intentionally.

There has often been a degree of indifference to children in human society. Let's not simplify or romanticise a past where endurance, not compassion, was often the governing imperative. Kids in pre-modern society were surely loved in a certain way. Yet, like everyone, they also had to earn their keep, play their part, or else … There wasn't much room for sentimentality in a society focused on survival. In an age of unprecedented affluence, however, adult self-absorption is surely a pointless and cruel form of harm.

Back to Sligo … The graveyard had a heavy air about it that suggested mass death. Interments too populous for headstones and a separate yard reserved for children. Wikipedia tells me that about 2000 people were buried in this graveyard. I sat for a while in the children's graveyard and wept a little. Doubtless like other parents in this situation, you can't help but imagine your own children in lifeless heaps under the surface. It's the only way to touch the awe of it all. From later reflection emerges a social sensibility; you begin to imagine the vast swathe of families cut down by The Great Hunger, or in Irish, An Gorta Mór.

This emotional ambush rocked my assuredness. I'd given this presentation many times in Australia and New Zealand, and now in Ireland. It was always well received. But suddenly, slumped sadly in a graveyard I'd not sought out, I wondered did I really know what I was talking about? What right had I to speak about harm to children in the kind of rich environment that these poor dead souls would have greatly desired?

Some critical reflection eventually reset my compass of thought to a slightly new course. And a sharper one, I think. First, I realised my own distant personal connection to where I sat. My ancestors left Ireland for Australia in the late 1840s in the wake of the hunger. This helped to ease the alienation of the moment. Second, I remembered that societal harm to children occurs both with conscious murderous purpose and unintentionally, but much more insidiously. The mass starvation of children during *An Gorta Mór* was murderous. But a more intentional evil was the butchering of Aboriginal children – usually by clubbing – around the same time by settlers and Native Police as related by Raymond Evans in his recent *A History of Queensland*.[7] It happened in other colonies too.

These graveside reflections helped me realise that it's right and necessary to work against both forms of harm. My work points to the way urban change is working against children at the insidious and unintentional end of the scale. In Australia today, the deep cultural roots of narcissism, and their consequences for children, are a potent form of this harm.

At the time of *An Gorta Mór*, the famous indifference of Victorian industrialists to children – for whom they were mere factory fodder – was giving way to a reformist view that sought to rescue childhood from industrialism. But the objects of this reform were English children, not the Irish, who, as colonised Catholics, were outside the humanist mind. English cultural critic Terry Eagleton writes that at the time of *An Gorta Mór*, England's esteemed economist Nassau Senior remarked that a million dead in Ireland would 'scarcely be enough to do much good'. This suggests a deranged mind. Imagine an economist writing this way today – it would certainly rouse us for a moment from our lethargic ignoring of kids' everyday wellbeing. Our indifference works much more quietly, below the surface, to erase children from the societal mindset.

A third idea occurred to me as I sat with the lost children. My own connection to famine was not merely ancestral. I'm part of an urban culture that is implicated in the hunger and death inflicted on children in developing countries today. A great food recession is sweeping the Earth as agricultural output is diverted from human stomachs to petrol tanks. This would not be occurring if cities, where the majority of humanity now lives, were less voracious users of oil. Our stubborn car dependency has made cities more polluted and unsafe for our kids. And it has made us dependent on a declining and costly resource, which has had disastrous consequences for children elsewhere. Like junkies, we're lunging at a new quick fix, biofuels, as our oil supplies dry up. There has been no questioning of the habit itself until very recently.

The destruction of forests and conversion of farmland to biofuel production have made food much more expensive and thus less easy to obtain in the developing world. In 2007 a senior UN official, Jean Ziegler, called biofuels a 'crime against humanity'. A year later UN Secretary-General Ban Ki-Moon called for a comprehensive rethink of the enthusiasm for biofuels.[8] Even the IMF is worried.

I'd travelled to Sligo Town from Dublin on a slow, crawling train – at times I could have walked beside it and not been left behind. The GFC hadn't yet hit and humbled the Irish tiger's roaring economy, the freewheeling love child of European neoliberalism. The government had bold plans for a major improvement of national rail systems, but it was also building vast new road networks. It was the sort of 'balanced' transport strategy that has been favoured in Australia for some years; outwardly promising a new 'sustainable' approach to mobility, but in reality cloaking a business-as-usual focus on road building and a continued neglect of public transport.

Travelling back to Dublin with famine on my mind, I thought how the motor car cult is yet another instance of the adult self-absorption that is quietly harming children. We surely owe the children of the Earth an urban lifestyle that does not deny them food. The car has given us, or some of us, a lot of freedom for a relatively short time. It is a marvellous but dangerous and costly technology. Globalisation is urging the motorisation of our species. This is simply not a tenable prospect, environmentally or socially. And we in the West, who unleashed the beast, have to rein it in and, while we're at it, admit that the dream was a destructive fiction. Our lust for biofuels, and therefore also for the precious food-producing lands they need, is grievous evidence of our addiction to the car. The 'peak oil' siren still struggles to dispel our long hallucinogenic dream of costless, boundless mobility.

Contemporary Ireland is, by European standards, a young nation. Australia, by broader Western comparison, is also a young nation, despite the mounting hysteria over population ageing. Both countries have a special obligation to think carefully about what development path will best meet children's needs. Rapidly expanding cities and towns need close attention. The problems plaguing the outer commuter belts of Dublin and Perth are remarkably similar. Their common roots lie in an economic system that walks backwards into the future. In any country, urban development that ignores children's wellbeing will in reality work insidiously against them. The cities will get richer and brighter, but will be poorer places for kids and their carers. Our feast will be their famine. A paradox at the heart of modernity.

Anne Manne senses 'something fundamentally flawed' in our contemporary social makeup. She writes, 'By any historical standards, our society is marked by a radical individualism obsessed with the self.'[9] It is a special perversity of the age that we should spend so much time before the mirror and yet not notice our spreading blemishes.

Manne points to the latest, and perhaps most destructive, phase of 'individuation'. The search for the self has always been cast as escape from primitive servitude. Capitalism threw aside the bonds and superstitions of feudalism. We moderns have dreamed of freedom in the centuries since. We have strived for separation from nature, liberation from kith and kin, and release from all the other grubby dependencies of human existence. If modernity, despite its declared secular nature, harboured any mysticism, it was this dream. And the potency of this guiding myth of modernisation cannot be denied. We shrugged our earthly bonds. We learned not to suffer innocence. And we threw down the Kingdom of Heaven.

LEARNING TO SEE
THE EARTH

... Bruise
violet and viridian a threat
of storms I could conduct with an index finger wet
from the cup, catching a hint of what God
felt, trying for this, then that; learning to see the earth
as it is from failed experiments – even those we give
our hearts to and can't forget.

David Malouf, 'Like Our First Paintbox' (2007)

IN EARLY 2009 KEVIN RUDD DECLARED 'the end of neoliberalism'.
The defrocked priests (and priestesses) squawked that it could not be
true. They are in part right. We have a long way to go to get to a safer
and humane alternative but we don't have much time to do it. What
must be done to bring us to safety? What myths must we set aside?
Inevitably, we face a time of difficult adjustment, needing to set aside
ways of life as well as many accepted ideas, even, as David Malouf
says, 'those we give our hearts to and can't forget'. We must have a

clearance of old and useless intellectual stock. Out must go our belief in the endless natural abundance of Terra Australis, along with our faith in technology and other painless miracles meant to save us from the claims of natural necessity.

We must learn to see the Earth, at last. We brought to the land the European modern heritage; a people 'enlightened' by their increasing separation from nature. From this rupture the rise of individualism – and the modern 'free' subject – was made possible. The irony of this presumed freedom was perhaps not lost on the people who came here in chains and irons originally, but their arrival nonetheless heralded the imposition of a culture of independence and rationality on an ancient landscape and its guardians. Stephen Dovers makes the argument that we have struggled, to put it mildly, to settle a landscape that we were not well equipped to comprehend and sympathise with. Does this explain our savagery towards its original owners? We may never have an adequate answer to that question.

If we never settled, then as far-flung moderns we may find ourselves now struggling with the finiteness of nature generally, and encountering a fierce and frightening set of rejections from a land injured and misunderstood. Our long, languid attempt to settle Australia may never be finished, at least on the terms we have used till now. A new dispensation will be forced on us by an increasingly bellicose climate and by the depletion – perhaps even collapse – of ecosystems, biodiversity and resource stocks.

But this may be cause for hope, not gloom. When we recognise that the Earth's systems aren't infinite, but are a set of always potentially nurturing possibilities that need time and space for renewal, we can leave the arguments about finite and infinite nature in our slipstream. Our role in the Earth system must be as stewards, well aware of our dependence on the continual turning of the big wheel of renewal. By attempting, unsuccessfully, to destroy Aboriginal peoples, we managed to disrupt the 'sacred nexus between people and place'[1] that they had established over millennia. We need to craft a new nexus between ourselves and the land. The original inhabitants have much to teach us, I believe, if we are prepared at last to listen with humility. This is what I take the idea of 'learning to see the Earth' to mean.

'Learning' is the right verb, because it evokes the openness, humility and self-awareness that we will need to cultivate as we pass through a period – who knows how long? – of great distress and upheaval in our social and political economic systems. The progressive opening of nature's faultlines is what will at last teach us the lesson. In what ways must we endure and in what form will we survive this time in the kiln of change? We cannot say for sure, but in the speculations and suggestions that follow in this part of the book I believe I'm edging towards the more hopeful of prescriptions. Some of those on offer see our species massively culled, even annihilated.

I have great faith in human resilience, but I don't deny the enormity of what lies in store. This will be like global war, and I'm far from the first to say this. To pass through to a saner, more resilient accommodation with nature we will need strong civic and public structures, including what I term a guardian state, as the means to a safer end. And our cities, in some ways the source of our problems, must be reconceived as the lifeboats to take us through the storms ahead ... until we see land, until we have learned to see the Earth for what it is: our offended friend, but ultimately our best and necessary companion.

7

A CRUMBLING EMPIRE

> The challenge is not how to shore up a crumbling empire with wave machines and global summits but to start thinking about how we are going to live through its fall, and what we can learn from its collapse.
>
> Paul Kingsworth, British environmentalist, 2009[1]

CLIMATE CHANGE AND ENERGY INSECURITY are real and present threats to the stability and sustainability of human society. The imminence, scale and speed of both threats appear to overwhelm the principal mitigation strategies on offer. In recent years the scientific consensus has consistently pulled forward the horizon for action to prevent grave climate harm. It now seems that we have 10 years or less to make massive cuts to carbon emissions to avoid ecological default. Jim Hansen, director of NASA's Goddard Institute for Space Studies, writes:

We have to stabilise emissions of carbon dioxide within a
decade ... [or] many things could become unstoppable ... we
cannot wait for new technologies like capturing emissions from
burning coal. We have to act with what we have.[2]

This is pretty hard for a country like Australia, which is putting so
much faith in the lure of clean coal. Hansen seems to think it's a ruse.

Then there are fatalists such as James Lovelock, the renowned
British scientist and environmentalist, who believes that it's too late
altogether to save what we have. Nearing the end of a long and
brilliantly lived life, Lovelock argues that humanity must pass through
a storm of disruptive change and threat that will change our species
utterly. His 2009 book, *The Vanishing Face of Gaia*, issues a 'final
warning' to humanity based on a lifetime of trailblazing inquiry into
the Earth's natural order, with insights and discoveries consistently
confirmed by scientific consensus.[3] (But often not before he suffered
the agonies of collegial scepticism and rebuke.)

Lovelock's guiding premise is that 'our world has changed forever
and we will have to adapt, and to more than climate change'.[4] By the
latter he means all the other harms we have inflicted on nature, or
'Gaia' as he terms the Earth's natural system. These injuries include
the worsening acidification of the globe's oceans, which is linked to
global warming, and which threatens a collapse in humanity's food
stocks. Ultimately, as a parent, and as a timid thinker, I simply can't
go with Lovelock to the final stages of his prediction. This sees humanity, in this very century, battered to the point of near extinction
and, at best, re-emerging from these crises as a changed and mortified
species, hopefully at one with Gaia and freed from delusions. But neither can I ignore the case he presents for the end of business as usual,
and for the inevitability of massive, and probably cataclysmic change.
His prescription for us is in large measure pragmatic, seeing the im-

mediate goal of humanity as survival with a minimum of suffering.

Lovelock is not so exceptional or fatalistic as to reject the call of Hansen and other leading scientific fora, such as the United Nations' Intergovernmental Panel on Climate Change (IPCC), for swift intervention to reduce global emissions and turn us back from the point of runaway change. Lovelock probably sees chaotic climate change as foretold, but clearly also thinks that anything we can do to buy time to develop adaptive strategies is worth it. It's fair to say that the scientific consensus still believes we have the opportunity to bring our climate back under control and prevent the worst of scenarios, which sees average temperatures rising three or four degrees. That scenario would reduce the proportion of the Earth's surface that could support life. Lovelock thinks it would reduce the liveable surface to a few 'lifeboat' regions, areas that are now the cooler extremes of the Earth.

When you compare these two perspectives, you see that the common ground is the lack of time and the need for very swift interventions to arrest and turn back amounts of emissions. We must do more than simply rein in growth. Humanity must work to *cool* the atmosphere it has cooked since the Industrial Revolution. The 'fire in the heavens' must be doused. This means removing carbon from the atmosphere, not slowing our addition to it. It's a galling prospect, given that all our mainstream environmental discussions and efforts so far have been about just slowing the growth of carbon emissions.

Most of the green measures we have before us, especially national carbon trading schemes, ignore the short timeframe and the scale of the adjustment we must make to keep some sort of grip on our climate. Then there is the question of the disappearing resources – oil, water, arable land, forests, fish stocks, biodiversity – that seem to be irreplaceable, despite our best efforts at technological substitution and improved management. When it comes to 'global heating', as

Lovelock terms it, technological fixes and market adjustments both have vulnerabilities and timeframes that may make them unviable. The dream of 'clean coal' shimmers on a dangerously distant horizon. Lovelock insists that all 'technical dreams' have implausible horizons.[5]

Some schemes, notably the switch to renewable energy, are nonetheless necessary for our long-term security and sustainability. Others – especially the deathly lures of nuclear energy and biofuels – fail on both these counts. Lovelock supports nuclear energy as a clean energy form but neglects its inescapable role in weaponry and militaristic ambition. He accepts that we are entering an era of climate-forced geopolitical instability, and possibly heightened warfare. How can we contemplate a massive expansion of the global nuclear industry at such a time? He retains some hope for spectacular 'geoengineering' interventions which would remove carbon from the atmosphere and save the day. These include various proposals for capturing carbon using diverse media such as the oceans, weathering rocks and vegetation. They are all worth further work, because this approach might turn out to be our last hope. But nothing convincing or workable has yet emerged from the labs and the time for massive action has arrived.

Another dead end strategy, at least in terms of climate, is population control, a favourite theme of some sections of the environmental movement and of others with less charitable intentions. In the longer term we must restrain our fertility to reduce our ecological load; but this is no immediate climate fix. As Lovelock puts it, 'No voluntary human act can reduce our numbers fast enough even to slow climate change.'[6] Everything seems to fail the time test, which raises the prospect of our species failing the exam too, as Lovelock suggests.

For the cities, the heartlands of *homo urbanis*, some scientists and green advocates have for decades been urging urban containment and the general curtailment of growth as a solution to rising greenhouse emissions. The 'compact city' has become the new international ideal, at least in Western nations, and has been taken up with varying degrees of enthusiasm by governments and planning authorities. In Australia, state and territory governments have embraced the ideal of consolidation without doing a lot to actually achieve it. A notable exception is the recent series of regional plans for southeast Queensland. These have reined in a long period of freewheeling, haphazard urban growth. But it's fair to say that the compaction dream is more honoured in the breach than in the observance. Clive Forster, an Australian urban geographer, speaks of a 'parallel universe' problem, where our plans project a sombre ideal for growth – the compact city – while our ordinary urban development processes continue in their mostly unrestrained and uncoordinated way.[7] But even if we had the resolve to pursue compaction, could it stave off the climate crisis? Or even make a major contribution to that end?

Sadly, science and experience tell us that established planning and design processes can't do much to mitigate the climate crisis. I will outline evidence in the next chapter which shows that the project of urban consolidation is another technological fix that won't work. It's a strategy that can't reduce the growth in emissions fast enough. Having cleaned this misty-eyed ambition off the lens, another much more compelling role emerges for planning: the equally immediate task of adapting to a climate and landscape already changed (and continuing to change) by climate shifts, and by the biospheric perturbations associated with this. And before you think I am in favour of more business-as-usual urban growth, I am most definitely not. Australia's large metropolitan regions are now pushing outwards into areas that will be stressed by climate change, and which we must preserve in order to maintain critical environmental services such as water, biodiversity and recreation. Further, we face serious 'diseconomies', to use that fabulously silly technical term, if unmanaged urban spread continues. These imperatives, not energy use mitigation, compel us to rein in our cities and large urban areas.

As an urban scholar, I'd love to think that simple improvement to city planning and design could stave off the environmental crisis. It can't. The only feasible strategy to meet the threat appears to be a massive and sudden decrease in consumption and a rationing of key resources, especially water, oil and energy. British environmental commentator George Monbiot agrees, recently suggesting that the looming world recession may in fact be the pathway out of peril if it forces down consumption in rich economies. The title of his 2008 book, *Bring On the Apocalypse*, reflects a rueful enthusiasm for the collapse of our destructive economic order because it might buy us some time.[8] But it's hard to imagine without horror the many lives and communities that will be ground down – indeed it's happening now – as the contradictions of neoliberal capitalism come home to roost at the global scale. And as usual, it's the developing world that's copping the brunt of the collapse. This isn't, however, to imply mere victimhood on their part; I sense that many ideas for an alternative and humane political economy will emerge from outside the West.

We face a time of threat akin to global war. The peril is grave but not insurmountable. And yet the West's energies and resources have been poured into the fight against a much more ghost-like threat: terrorism. Sir John Houghton, former head of the British Meteorological Bureau and senior member of the IPCC, has observed that climate change kills more people than terrorism and poses at least as great a threat to human security as 'chemical, nuclear or biological weapons, or indeed international terrorism'.[9]

Lovelock thinks that we should throw everything we can at mitigation, but only in the hope that it will earn us time to become better prepared for the holocaust ahead. Inevitably, Gaia will restore the natural balance by culling our species through transition to the sort of hot climate regime that pre-dated our emergence. I hope there is another way to safety, even if it means pain and adjustment. His strategy hinges on two changes that might open an exit door for us: a final and thoroughgoing rejection of Prometheanism, the suicide cult of modernity, and a radical and immediate cut in consumption of the Earth's capital. Do not discount the extent to which our political ideologies, economic system and globalised cultures will refuse this exit, because it means letting go of the bright dream of technological escape and the warm mirage of business-as-usual thinking. None of our embedded political and economic systems, or the ways of thinking that surround them, will go willingly.

But go they must. First, we must tear away the grip of our self-defeating thinking, without embracing the many dead-end or misanthropic ideas that have masqueraded as alternatives. Deep green approaches that are hostile to human wellbeing are one such cul-de-sac. We are, as Lovelock, points out, an integral part of Gaia; we just need to find a role that secures our long-term survival in an ecological system that is showing us that it will not tolerate our current approach. There must be a fire sale of failed philosophies. This means re-evaluating market forms that drive us ineluctably towards overproduction. The broad contours of an alternative are given some thought in Part 3.

What does this ultimately mean? Do we need to succumb to a miserable vision? Must humanity resign itself to the frugal living that our forebears did all they could to escape, including through (Industrial) revolutionary means? Again, Lovelock is helpful in confronting this deadlock. He tells us that the Earth is neither finite nor infinite, but a natural system that is capable of renewing itself if we learn to live in sympathy with its needs. If we were to do so, it would open up the prospect of a naturally aligned civilisation that could grow in all sorts of powerful ways – culturally, spiritually, socially, even materially – if committed to the imperative of renewal. We are a long way from these shores. As Lovelock says, 'Our [continuing] error is to take more than the Earth renews.'[10]

As part of our learning we must, I think, smell the soil of privation for a while. As I've argued above, there seems no way that we can prevent a long period of massive consumption cuts if we are to rescue the climate, our resource base, and ultimately our relationship with Gaia, the mothership. We've some making up to do.

So the second hinge in our escape hatch is the necessity of rationing. I'm convinced that we will eventually realise, I'm hoping not too late, that we have to cut and cap consumption of the things that are harming our environment and the resources which are in short supply. This will mean much more than paying a bit more for electricity. It could go as far as taking decisions about harmful consumption out of individual hands and imposing systems of safe and fair rationing that do not exceed ecological limits. Not a pleasing prospect and one that has dangers for democracy, if the imposition is centralised and becomes autocracy. The secret will be to maintain fairness and solidarity as governing principles of resource distribution and use.

This idea goes against the political economic system that has governed the West, and increasingly the globe, for two centuries – the market order. Rationing means a massive and centrally coordinated cut in consumption. It therefore confronts overproduction, and a system that, given its will, would keep increasing consumption, at least of material, exchangeable things. We desperately need an alternative to an economic order that is hardwired for growth, in the narrowest sense. My colleague Nicholas Low talks of the need to 'suspend capitalism' for a period, as we did during the last global conflict, until order can be restored. But I believe that the transition to scarcity and rationing must mark the passing of the market order, at least as we have known it, not just its temporary suspension. In the last global conflict, Allied societies were rallied and joined to a 'just' cause by the promise of something different in peace, if it were won. This time I think we should look to something really new, something that releases us from overproduction.

Australia's great urban commentator, Hugh Stretton, likens the sustainability crisis to a time of war: the only path to salvation lies in a swift, centrally coordinated response, including resource rationing and the outlawing of some forms of consumption. Cities will have to learn to ration themselves. Stretton's most recent book, *Australia Fair*,[11] addresses the environmental menace and concludes, as I do, that some form of resource rationing will be forced upon us. Reflecting on the great rationing exercises that saw us through World War II and its reconstruction phase, Stretton believes that their success can be repeated:

> But it is likely to depend, now as then, on three achievements which look unlikely as this is written. We must believe the dangers are real and deadly. We must hope to survive them by radical action, self-restraint and sacrifice. And we must attract the necessary solidarity by a serious reduction of our inequalities.[12]

Restraint, sacrifice, solidarity: these are words that could shatter the neoliberal dream. Generating large consumption cuts is surely the province of the nation state. By electing the Rudd government in 2007, Australians authorised a serious alternative to Howard's strange mix of dithering and denial. So the task of reining in consumption is now theoretically possible. The subsequent Garnaut Climate Change Review produced a sharp call for structural change and restraint. But in 2009 this appeared to have been submerged in the messy politics of the government's tellingly wordy Carbon Pollution Reduction Scheme. The parliamentary brouhaha revealed, amongst other things, a deep vein of climate scepticism in the conservative parties. Against this eco-cidal view, the government presented a much more reasonable, 'responsible' program of adjustment. The Opposition rallied in that year, though messily, and countered with its own, 'better' scheme that was kinder to polluters. And so it went. Meanwhile, the heavens burned ...

We may be entering a more sophisticated phase of Prometheanism, based on the belief that resource use can be decoupled from economic expansion. The obsessive focus on market instruments – massive as emissions trading schemes may be – will be viewed in hindsight, I think, as another thoughtless delay on the road to an inevitable and difficult confrontation with natural necessity. The need for consumption cuts will, before too long, shock any system wedded to overproduction.

So where does this bring us now? I think the West is experiencing a moment of rude rousing from many dreams. In Australia, it's as if the whole dormitory has woken simultaneously, angry nature shattering the windows with sudden force. What happened to the warm narcosis of the miracle economy? How did the dream of freedom morph into a schlockbuster about global warming and oil vulnerability? Who brought this horror upon us? As consensus strengthens on the threat posed by global warming, the tendency to blame and punish also strengthens, as John Howard found out. For some time, a swathe of the urban commentariat in Australia has been blaming the people who apparently supported Howard's inaction on climate change. They have a seductively neat answer to the question of culpability: suburbia. Suburbia is the consumptive beast whose appetite has ruined us all. The next chapter will oppose this answer, arguing that it's our 'vortex cities' and our failed political economy, not simply our suburbs, that explain the crisis.

Maintaining equity, and therefore solidarity, will be critical to the success of mitigation and adaptation strategies. We must work out how equity is to be maintained in the face of threat, disturbance and displacement. As the author Steve Biddulph put it, 'we co-operate or die'.[13] Yet co-operation will not thrive without a fair distribution of burden and effort. It is, as Australian social historian Mark Peel states, time to 'talk of shared sacrifices led by those with most to give'.[14]

A prescient letter to the editor in Brisbane's *Courier Mail* in late 2007 captured the essence of the problem. The letter laments the role of elites in climate debates, including the 'knowledgeable' and the 'rich and famous', observing their continued ability to 'fly their private, corporate or government-funded jets' while the 'numerous' rest 'are warned that if the worst does materialise it will be our fault if we do not immediately turn off our air conditioners, use smaller cars, burn less fuel and save all the rainwater we can muster'. Moreover, 'no rationing is mentioned to bring equitable sharing of the load. So the haves can still outspend the have-nots.' Keep the theme of rationing in mind; as things begin to turn for the harder, it will be discussed and argued about more and more. The next chapter provides a meditation on this challenging theme.

I think the letter's sentiment anticipates a divisive politics that may well emerge when the broader community begins to comprehend that it is consumption by the wealthy that is most environmentally harmful. The poor(er), as ever, are not the core of the problem. While a small part of me cheers the justice of the claim, the outbreak of suburban complaint mostly fills me with dread. We cannot afford yet another obstacle in our preparation for what lies ahead. There is an immediate need to promote, and if necessary prescribe, a culture of moderation amongst elites. This will include restraints on the most conspicuously damaging forms of consumption, including air travel and the import of bulky and weighty luxuries.

I feel certain that another crippling drought or resource crisis is on the near horizon. Sadly, it may be what is needed to shift us to the war footing that Stretton and many other commentators urge. The big question is, how do we make the transition to a guardian state in the wake of decades of neoliberalism? Who even knows any more how to steer things for the public good when it comes to primary resource conservation and rationing? We must begin to talk about this guardian state, especially about how it can be reconciled with an unyielding commitment to democracy and fairness.

The latest human dream may be fading, but its political and moral legacies will not easily be overcome. Most of our leadership cannot think outside the narrow frame of neoliberalism. The universities have lost, or had removed, much of their capacity to craft alternative thinking. We are suffering serious ethical and intellectual deprivation at a time of peril. It's been a long time since we had faith in the kind of concerted public endeavour that will be needed. Social justice and social solidarity seem quaintly archaic ideals, and the concepts of restraint and modesty seem barmy. These will all have to be restored to public life and to institutional purpose if we are to find a safe passage through the coming storm.

There is nowhere we can run to. We must find and make a new Australia here. To begin this restoration, we must rouse ourselves from the modernist dream and reawaken our most basic human obligations – to each other, to those to come, and to the ecology that will nurture (or at least outlast) us all. Without these commitments, society might survive but democracy may not. At worst, nature may simply decide to go on without us. In an age of ambiguity, we can be sure of one thing: *homo urbanis* will meet its destiny in the cities. These, I believe, will be the real lifeboats of humanity.

8

THE URBAN VORTEX

> Avoiding dangerous climate change is impossible – dangerous climate change is already here. The question is, can we avoid *catastrophic* climate change?
>
> David King, UK Chief Scientist, 2007[1]

THE AUSTRALIAN CITY TODAY – take any state capital – is a vast resource use system, dependent upon ever more extensive supply chains and regions. Economic globalisation has extended its consumption tentacles further and further across the Earth. When Sydney catches an economic cold, as it has in recent years, it hurts all sorts of far-flung suppliers and communities. Globalisation has increased the interdependency of the world's regions, very often meaning that people in poorer rural areas – and indeed people in cities too – rely for survival on cities they will only ever see on television screens.

Urban systems have always driven and amplified the core processes that harm climates and exhaust resources, especially during the recent freewheeling globalisation which saw the plunder of the

Earth's resources and – in many places – the weakening of environ-mental and civic safeguards. And yet an anti-urban perspective would be self-defeating. Humanity has embraced urban life and isn't likely to favour alternatives, whatever they may be. Opposing urbanisation or even urban life won't get the cause of change very far. It would be no more useful than shouting into the wind.

Cities and large settlements are where the world, and especially Australia, must confront and deal with the problem of environmental bankruptcy. They are sites for containment and resolution of the threats we face, as much as they are the sources of many problems. The goal of sustainability requires transition from urban vulnerability to urban resilience. We must learn to see the urban vortex for what it is, a set of countervailing possibilities; a wellspring of harm, and a field of possibility; a whirlpool of failing ambition and an island of refuge and renewal.

If human survival depends upon us learning to see the Earth, and to recognise its claims on us, what is to be done with cities that have their heels on the throat of nature? The point is particular, not universal: most urban areas in the developing world use and despoil far few resources than even the best in the West. It is the developed world that must begin the transition to a new natural order by bringing into reality a peaceful, sustainable and equitable urban model. We have to get our houses in order – literally – before we can expect the poorer, more insecure world to follow. And we have to open our minds to the idea that we have much to learn about resource stewardship and human care from the developing world.

The first tasks, as I emphasised in the previous chapters, are to secure what we have and prevent the realisation of the worst threats from warming, resource depletion and economic breakdown. This necessitates a commitment to equity and solidarity and the adoption of new civic virtues that reflect this. We must pass through a storm of change, and for most of us this means entering – to adapt Australian geographer Phil McManus's term – an urban vortex from which our cities and communities will emerge transformed and strengthened. That's the goal we must set ourselves. What does this mean for urban Australia, whose cities I cast earlier as on the edge of a series of environmental and social defaults?

The principal feature of contemporary urban Australia is the overwhelming significance of its suburban landscapes. Donald Horne famously described us as the 'first suburban nation' in his groundbreaking 1964 work, *The Lucky Country*.[2] A majority of urban citizens live in some form of suburban setting. Most people in sea-change, even tree-change, areas live in suburban estates of varying densities. This simple social fact determines not only the conditions for planning, and urban policy generally; it must also be the starting point for the national effort required to respond to climate change and oil vulnerability.

A further consideration is urban social equity – this is made more necessary than ever by the global economic crisis and the huge social stresses it continues to generate. At the larger urban scale, Australian cities are marked by a significant wealth divide between lower-density suburbia and the inner-urban higher-density domains. This is a general divide – there are concentrations of poverty and wealth in both places. The most entrenched and worsening areas of disadvantage are, however, in suburbia, particularly within the ageing middle rings of the major metropolitan regions. It would be risky for public policy responses to environmental pressures to ignore this uneven geography of advantage. It would be inequitable and self-defeating if policy included measures that unfairly burden suburban Australia.

A widely shared, though not unanimous, assumption in scholarly and policy circles is that suburbs are at once the source and the worst reflection of the sustainability crisis. This view has resonated with increasing strength in 'expert' urban commentary. Imported US terms, such as 'sprawl', 'smart growth' and 'new urbanism', have begun to dot the landscape of Australian urban scholarship and debate. Geographer Clive Forster recalls comment from a national radio documentary in the early 1990s:

> Australian cities have reached a mid-life crisis. Two hundred years after European invasion and the beginnings of urban development in this country, we are looking down at the sprawling belly of our cities and exclaiming, 'Oh my God, how did that happen?'[3]

Complaining about suburbia has a long history, especially in the arts and in parts of the media. In 1964 Donald Horne noted how 'bohemians and rebels attack "suburbanism"'.[4] But this criticism now has a virulent green edge, adding ecological waste to the schedule of suburban crimes. Waste and sprawl are considered synonymous.

Although 'sprawl' is correctly defined as 'unplanned low-density urban development', the term has tended to be used for all of suburbia, well planned or otherwise. It has been granted deathly potency in scholarship and commentary, especially in the United States. Joel Hirschhorn's 2005 book *Sprawl Kills* reports that sprawl annihilates comprehensively by also stealing 'your time, health and money'.[5] Australian architectural critic Elizabeth Farrelly provides forensic detail: 'the traffic jams and the water shortages, the poisonous air and the childhood asthma, the obesity, the neuroses, the depression'.[6] However, most suburban Australians remain unaware of, or untroubled by, the sprawl bogey.

Green critique of sprawl is now the leading kind of censure of suburbia. It has been bolstered by studies in Australia and overseas that have concluded that low-density urbanism is wasteful and polluting. The international chorus of critique fed by this work has neglected the very different suburban forms that manifested in developed nations during the 20th century.

The criticism of suburbia assumed by much urban commentary has poor scientific foundations. In particular, recent Australian analysis points to the consumptive lifestyle, not the nature of one's dwelling, as the root of environmental woes. The 2007 urban consumption analyses produced for the Australian Conservation Foundation (ACF) by the Centre for Integrated Sustainability Analysis at the University of Sydney turn conventional eco-criticism of suburbia on its head. The Main Findings report concludes that:

> despite the lower environmental impacts associated with less car use, inner city households outstrip the rest of Australia in every other category of consumption. Even in the area of housing, the opportunities for relatively efficient, compact living appear to be overwhelmed by the energy and water demands of modern urban living, such as air conditioning, spa baths, down lighting and luxury electronics and appliances, as well as by a higher proportion of individuals living alone or in small households. In each state and territory, the centre of the capital city is the area with the highest environmental impacts, followed by the inner suburban areas.[7]

The point is that total household energy consumption, and therefore greenhouse emissions, are made up of both *direct* and *indirect* components. The former is the energy used to maintain everyday lives – petrol, gas and electricity – and the latter is the energy embodied in goods and services consumed. While most, if not all, of the focus of urban commentary and policy is on direct energy use, in reality, 'direct household and person use accounts for only 30 per cent of our total greenhouse gas pollution, 23 per cent of our total water use, and just 10 per cent of our total eco-footprint'.[8]

So the energy use most influenced by the size of our house and its location only accounts for a small share of greenhouse emissions. Free-ranging consumption of goods and services produced well outside our life-worlds is causing the problem. Shoving everyone into highrises won't solve it. In fact, if every Australian household switched to renewable energy and stopped driving their cars tomorrow, total household emissions would decline by only about 18 per cent. It's our consumption appetites that are the real problem – and the biggest guts are not in suburbs.

The 'suburban gothic' tale has produced its equally melodramatic counter-narrative, the Great Australian Dream Swindle. The Dreamers are led by the pro-market Institute of Public Affairs (IPA)[9] and various development lobbies, abetted by foreign consultants. They are joined in a crusade, urging the masses (apparently) behind them to push the suburban frontier ever outwards. The crusaders erect their own bogeys: planning, environmentalism, the blabberland of urban critique.

This tale bemoans a stolen generation of home ownership dreams. We see a cinemascope fable of hopeful newlyweds in wagons turned back from suburban frontiers by unfeeling black-robed bureaucrats. The black robes have halted the natural order of suburban things by slowing the tide of brick veneer. Those who weave this tale wish to safeguard the long slumber of suburban conventional wisdom. Here, social intelligence is reduced to the pragmatic axiom: what has (appeared) to work will always work, and therefore must always be right. Not the attitude a society needs to survive the threat of ecological collapse, I would suggest.

Neither set of protagonists, Goths or Dreamers, comprehends the sustainability threat. Global warming and oil vulnerability can't be ignored, and neither can they be solved through simple manipulations of urban form such as densification. I think both debating positions are hopelessly utopian and, if I may say so, contrived: one reduces nature to a one-way street where physical form determines human behaviour; the other dismisses nature as a frontier land for infinite exploitation. Both testify to the power of arguments based on distorted views of nature, and both produce collective dreaming and inaction.

My alternative imagining of the urban takes nature as a starting and ending point for human experience, including urban life – not something from which we can free ourselves or which we can attain mastery over. Nature can't somehow be placed beyond or behind an urban growth boundary and then exploited or protected; it shapes (rather than determines) and makes possible the entire urban experience, including life in the gilded towers of apartment land. The ACF consumption atlases remind us of both how much our inner cities consume, and their dependency on nature. They signpost a way through the debating polarities that have been blocking movement towards a more fluid conception of the suburbs. This conception sees them as shifting landscapes of social and environmental possibilities; neither dystopias nor utopias, but human life-worlds whose physical qualities inform but do not determine their sustainability. This view refuses to fix suburbia as a landscape with a preset natural disposition that cannot be changed.

One simple way to falsify this view, if you live in suburbia, is to talk to an elderly neighbour who has lived in their home for a long time. Discussions like this reveal a suburbia that was immensely more sustainable than the model we have now, and that was based much more on modest consumption, mutual help and local provision. I think of the suburb where I live in Brisbane, which was developed largely after World War II and contained much war service housing. You can see the suburban modesty in the tiny garages, the large lots where vegetables were grown and chooks were kept, the many small and handy shopping strips (some still thriving, I'm happy to say). A farmer still does rounds with a produce truck, bringing food to front doors. The houses were sturdy weatherboard constructions that weren't air conditioned. These homes were vastly less consumptive than the inner cities of today. The historical analyses of urban scholars Patrick Mullins and Patrick Troy show us a suburban form that provided for much of its own needs.[10]

Even if the ACF's analyses are flawed (other scientific evidence backs them up), there's another reason why compaction isn't a means of escape from climate peril and resource depletion. There's simply no time. Urban environments are a highly fixed form of capital, and they take a surprisingly long time to reconfigure to the degree envisaged by some consolidationists – usually generations. As I argued in the previous chapter, the only feasible strategy to meet the threat appears to be a massive and sudden decrease in consumption and a rationing of key resources, especially water, oil and energy. The GFC has won us a little time, but we must work swiftly now to design and construct an urban order that can deal with resource rationing and scarcity without descending into a Hobbesian 'war of all against all'.

In 1970, Hugh Stretton, Australia's first great urbanist, wrote and self-published an urban bestseller, *Ideas for Australian Cities*.[11] The book disdained the anti-suburbanism of elites and offered a much more intelligent assessment of suburbia's strengths, weaknesses and possibilities. Stretton's scholarship enormously influenced generations of urban professionals by setting the compass towards equity as an ambition for planning and urban policy generally. It was sourced in a deep historical tradition of writing and activism that sought to correct the deadly injustices that had emerged in the early cities of capitalist industrialism.

Planning itself was born into a litter of reform movements in England (and later its colonies) that strove to replace the murderous urban inequities of laissez-fare industrialism. It joined battle on the fields of health and housing to correct the ills of early urbanism – the disease, environmental blight and social chaos that, combined, threatened the entirety of Victorian society. Working-class anger flared and reformists worked, knowingly or not, to douse the fires of revolution. The cause of urban justice was deeply inscribed in the 20th-century welfare state and its commitment to universal service provision, not governed by market power. It was a creaky commitment, to be sure, but one that drove the establishment of public housing, town planning and, eventually, a form of national health (Medicare) in Australia.

But recognition of and commitment to this urban ideal has waned. I doubt planning students at university, for example, ever hear the word 'equity' uttered in the course of their studies. These are the professionals who will have to help guide us through the stresses and breakdowns to come. They need to know how the ideal of fairness can be applied to the complex task of adapting urban systems in the face of change. More broadly, the language of justice needs to be restored to urban conversations generally if we are to find common purpose and strength in the midst of what will surely be a set of divisive crises. Our responses to threats and crises need to be steeped in social fairness, not electoral sensitivities, and to beat back countervailing considerations such as competitiveness and 'cost neutrality'.

There are terrible scales of natural disruption facing our cities. Entire water, energy and even urban climate systems are crumbling under pressure from antagonised natural systems. We cannot any longer avoid culpability for the damage we have caused. The scales of natural justice contrast starkly with the fragile institutional systems and meagre resources dedicated to urban resilience. In one sense, planning is the urban Maginot Line, and its capacities have already been sidelined by the global environmental menace. The ACF work indicates how moderate an influence urban form – the main object of planning effort – has on energy use and greenhouse emissions. However, urban form has a much more potent bearing on household use of, and dependency on, the direct forms of energy – notably oil – that are likely soon to be in scarce supply.

In the fight against global warming, I see planning's prime contribution as urban adaptation in search of climate-resilient cities. Good planning and design can reduce the vulnerability of cities to shortages in key resources – water, coal and oil. An immediate and wholesale improvement to public transport in the suburbs is the first planning improvement we should make in our climb towards climate resilience. Other insights and possibilities will emerge as we free ourselves from a debate about the suburbs that is polarised between censure and celebration.

When we reinstate history to urban discussion, we recover the alternative suburban futures discarded by both the Goths and the Dreamers. Outright proscription and simple prescription give way to consideration of new possibilities based on old insights – in this case, perhaps, a suburbia that recovers the values of modesty, solidarity and connectedness, but in new ways. At present, as Patrick Troy points out, the censure of suburbia is blocking thought on alternative possibilities, including the prospect that it may be the landscape best suited to safe adaptation in a warming climate.[12]

The suburbs' space and greenery offer immediate resources for onsite collection and disposal of water, generation of energy and production of food. Suburbia's adaptive potential has been understated or ignored by commentary and policy. Others are pointing this out to us. In 2008, renowned international ecologist Herbert Girardet told the International Solar Cities Congress in Adelaide: 'The suburb is perfect for low energy ... Low density is good for wind and solar power because there's more space to generate locally.'[13]

Scorn of suburbia weakens more than our ability to think of a way out of the looming crisis; it also threatens solidarity by demonising the social mainstream. The Australian scholar Aidan Davison argues that anti-suburbanism engenders disenchantment and withdrawal by the (sub)urban civil society that originally gave birth to environmentalism.[14] The hostility thus aroused is hindering the generation of a societal response to global warming and oil depletion; it fails Hugh Stretton's tests by unfairly apportioning blame and by undermining the conditions that produce solidarity.

The suburbs will be the main theatres in the defensive war against global warming, and need to be engaged and treated fairly in the debates and actions that will address climate change and energy insecurity. The first great task of urban adaptation must be a green suburban renovation. A critical view of the urban environments created by what Clive Hamilton has termed the 'growth fetish' economy of recent neoliberalism is the first step.[15] These landscapes include the walled estates of civic refusal that pepper our cities, the more environmentally egregious mega-homes, and the vertical sprawl produced by wild, market-driven consolidation. The narcissism of communal gating can't be allowed to continue if we are to rebuild the solidarity needed to confront the stresses ahead. Equally, we must restore the material capacity and civic confidence of those in these suburban exclusion zones.

The climate emergency presents the most compelling challenge for policy and science committed to urban resilience. Maintaining equity and therefore solidarity will be critical to the success of mitigation and adaptation strategies. These values will need to be reinstated politically to safeguard the social order in a long period of stress. Urban science and policy should ponder how equity is to be maintained in the face of threat, disturbance and displacement. In particular, if resource rationing is required, how is fairness to be maintained? This question will be taken up in the next chapters.

In the wake of global economic downturn, we assume that aspirationalism is a deflated and discredited project. I'm not so sure. Australia is hopeful of economic recovery. Consumer confidence surveys are prayerfully fanning the flame of consumption. Production systems maintain their mindless bonfire of nature. Everything is waiting to arc up again. The 'eco-rat' commentariat is waiting to refire the bellows of neoliberal puffery. The sceptics will cloak their foolishness in 'science'. They will urge a great effort at doing nothing again. The 'leave it to the márket' refrain will be loudly chorused again. As we pass into the vortex of change, do we have the will to shape our own destiny? Can we rouse from our slumber and take command of our cities, the vessels that will carry most of us through what is to come? It's the question of an urban age.

9

THE GUARDIAN STATE

> I need a political answer. This is an emergency and for
> emergency situations we need emergency action.
>
> Ban Ki-Moon, UN Secretary-General, 10 November 2007[1]

MY ARGUMENT TO NOW HAS BEEN that humanity, and all that
depends on us, is imperilled by a series of social and ecological crises
of our own making. And when I say 'all that depends upon us' I mean
the Earth – at least all its living systems – which we now have the
capacity to drag down with us if we fall. James Lovelock does not share
this view; he believes that Gaia will prevail, by withdrawing its support
for us and pulling other forms of life back into a new equilibrium that
will go on without us. Surely he cannot, however, discount the terrible
destruction to all forms of life that even this 'hopeful' scenario must
entail. It all seems too appalling to contemplate, but this is the age
when we must resolutely think the unthinkable.

My version of hope is distinguished from Lovelock's for two reasons. First, because as the collapse theorists, such as Jared Diamond,[2]

have argued, we have arrived at a point in our species' development where we can for the first time think 'reflexively' about survival: that is to say, learn from the mistakes of our forebear civilisations, which led to their destruction, through reasoned historical reflection.

If I've been a little hard on 'modernity' in the pages to now, let me start to pull back from this assessment with a serious qualification. I will say some words in defence of the Enlightenment and of modernisation in general. My position is to be critical of the excessive and debased forms of these great liberating projects that we have adopted under the cover of industrial and free-market capitalism.

German sociologist Ulrich Beck reminds us that the move from magical to modern thought was based on two guiding values: reason and doubt. *In dubito ergo sum*, to complete Descartes. Beck describes how the rise of Prometheanism in early industrial capitalism corrupted the Enlightenment project, producing 'excessive rationalisation' in society and politics, and ignoring doubt.

The great migration of humanity from rural to urban settings that was once regarded as the harbinger of development and democracy is increasingly coupled with global ecological collapse. The destruction of habitats for urban development and the insatiable need for materials for urban living continue unabated as the shift to cities goes on. It is certainly not being managed in the interests of an increasingly intemperate nature. In less than a century, we have become an urban species, *homo urbanis*. From nature's perspective, *homo diabolos* might be a more apt name. More than half of humanity now lives in large urban settings. In Australia, just about everyone does.

Our challenge is to sever ourselves from this sinking project before it takes us down with it. But this is not to urge retreat to the twilight from which our forebears fled. We must renew and redeploy the doubt-honed faculties that were liberated by modernisation to identify the tendencies for self-harm that threaten our contemporary civilisation. This thought renewal must, as I have argued, begin with us learning to see the Earth again. Prometheanism, and all the self-denial that went with it, was a most unnecessary debasement of an enlightened human mind. Indeed, it was the very negation of liberation of thought. The crumbling of the global empire indicts free-market capitalism and its crises of overproduction – this time in the form of an ecological crisis that imperils our continued existence.

This is, of course, the point where many readers will leave me. The pragmatists will say, 'Yes, but what's the point of structural critique? The alternative socialist projects of the 20th century have failed or withered to the point of irrelevance.' This suggests that US commentator Francis Fukuyama was right: capitalist liberal democracy is the (dead) 'end of history', a suburban shopping mall with no exits.[3] Many who hope for a better alternative witnessed the passing of autocratic socialism with relief, not regret, though the achievements of those societies bear as much examination as do their failings. Amongst their great faults was imitation of Western Prometheanism ... albeit with different consequences. The Soviet empire left poisonously polluted sites and rising CO_2 emissions, but the lack of a commodity-based consumption system restrained its overall environmental footprint. And now memory of these failed utopias fades. Do not discount the extent to which the recent GFC, and the unfolding natural catastrophes, will drag the 'winning' system back into the court of popular opinion – and very possibly censure.

My second cause for optimism is that we have recently shown ourselves capable of resolute species-level action that saw off a major (to put it mildly) environmental threat. This was the huge and concerted effort that went into combating rising CFC (chlorofluorocarbon) emissions in defence of the planetary ozone layer on whose continued existence we depend. The amazingly accomplished Lovelock actually invented a measurement instrument that helped bring our attention to the mounting CFC crisis. But while he seems to not make too much of the success of the world community's response, I believe it signals our potential to see off or at least contain the even greater threat of global heating. The CFC response involved some pretty strong action, against the currents of neoliberal globalisation, by governments united in fear by what science was telling them about the depletion of ozone. Unfortunately, the equally compelling evidence on climate change has had to deal with much more entrenched denial and inaction.

This might be as far as my optimism goes for now. I am not pleased by my belief that resolve and action will probably not happen until we are in the grip of a manifest emergency. The current 'clear and present danger' hasn't been enough to stimulate a genuine response to climate warming and resource insecurity. I think we will wait until a series of natural shocks wakens us finally from our dream and pushes us, perhaps too swiftly, into the next world. Let's hope we have some time to shape what that world will be like.

If the empire must crumble, what must we hope it gives way to? My thoughts on this are broadly outlined in the last part of the book, which imagines a more secure space for our civilisation in Gaia. That's a long way off, because we must first pass through a period of turbulence that will stress our species terribly. As we learn to see the Earth again, we will surely have our faces rubbed in the dirt of our errors for quite a while. Privation, disorder and stress will confront us in a variety of forms, some of them terrible. Cities will face water crises and weather extremes, there will be rising argument over scarce and costly resources, and whole ways of life, especially in rural Australia, will fade to dust. To counter these threats, we must commit ourselves to restraint, sacrifice and solidarity.

It's already obvious that our climate is heating up. Average temperatures are rising across Australia. A recent study by some of our best research institutions, the Bureau of Meteorology and the CSIRO, confirmed that the drought that has gripped southeastern Australia for the past decade or more is the consequence of a permanent climate shift, not part of a 'normal' long-run cycle.[4] There is science also to demonstrate the permanent loss of rainfall and catchment capacity in Perth and the southwest of Western Australia. This is reducing the productivity of food bowl regions, especially but not only the Murray-Darling Basin, and confounding the planning we have done around water and agriculture for many decades.

In response, the Victorian Farmers Federation President spoke airily of 'doomsday people in climate change' and his preference for scepticism, but admitted that hope, like the climate, was running dry in the rural sector.[5] These are the people who must feed us, perhaps until we have learned to see the productive Earth in our own metropolitan regions again. Our major cities once sourced much of their food in their immediate hinterlands, and we must commit to the restoration of this, as the clamps of shortage start to tighten.

Cities in drier parts of the world, from Los Angeles to Athens to Melbourne, seem suddenly more vulnerable than ever before to massively destructive fires originating in their peri-urban edges. Our water systems are under pressure, gas and oil reserves are steadily declining, and biodiversity and fish stocks are plummeting. The shift is underway. It will confront and ultimately negate the political and economic foundations of the globalised neoliberal order, and thus a way of innovating, producing and consuming that has been in place since the Industrial Revolution. Paul Kingsworth summarises the unfolding situation:

> The writing is on the wall for industrial society, and no amount
> of ethical shopping or determined protesting is going to change
> that now. Take a civilisation built on the myth of human
> exceptionalism and a deeply embedded cultural attitude to
> 'nature'; add a blind belief in technological and material
> progress; then fuel the whole thing with a power source that is
> discovered to be disastrously destructive. What do you get? We
> are starting to find out.[6]

Yes, so what happens as the crumbling of this empire advances? I don't have a crystal ball, but I think we in Australia have to contemplate this prospect now, probably before most other parts of the 'developed' world have to. I spent a good deal of time in Ireland last year, where in many quarters (but not the scientific community) the prospect of global warming is still being received with whimsical relief. If we had their climate we'd probably feel the same. The rolling dust clouds of change, of drought, of capricious weather, haven't started to bite in such regions.

So, we in Australia might be amongst the first Western peoples to be ambushed by global heating. We may well join parts of the developing world, especially in Africa and the low-lying southern 'mega cities', on the frontlines of climate change.

How are we going to respond to what our national, state and local politics may describe as environmental ambushes? Of course they will be no such thing, given the long period of warning from our scientific community and global science generally. I am certain that we will look back on the denialism of the Howard era as pure perfidy. But now we need to go much further than electing governments, like Rudd's, that 'accept the gravity' of the 'climate challenge'. We need to construct a state that is capable of managing the massive stresses that climate change will generate. How will our nation, our communities, manage these disruptions?

I propose that we will need to reshape government in the form of a 'guardian state' that can, like Allied governments during World War II, guide us through a time when 'normal' social and economic systems must be suspended, perhaps even abolished. Such a state must embody a kind of resolve that our governments don't presently possess. Its foundational premise is the emergency facing us, and its guiding values must be the social trinity I introduced in the previous chapter: restraint, sacrifice, solidarity. Importantly, the creation of this state means the suspension of capitalism, at least as we have known it. This is not as radical as it sounds. It was just so in World War II, when Allied states suspended markets and many other things, including some civil rights. Wholesale freezing of 'the system' was *democratically* possible. The sense of war emergency was reinforced by the popular disappointment in the functioning of the market system resulting from the Great Depression.

Some of our most senior scholars invoke the memory of global war to explain the scale of changes needed to bring us safely through the warming crisis. As reported in Chapter 7, Hugh Stretton believes that wartime resource rationing will be needed to rein in consumption. Sociologist Michael Pusey argues that 'a constructive adaptation to global warming will give rise to structural and cultural changes of a kind that we last saw in the aftermath of World War II'.[7] Pusey knows that the state response necessitated by the crisis will decisively end our three-decade experiment with neoliberalism and bring the greedier beneficiaries to heel:

> So also global warming might give us an opportunity to mobilise power in a way that brings vested interests to heel. More fundamentally, it has the potential to restore the legitimacy of state intervention and to generate the needed cultural energy for nation-building government – those very

resources that our economic reformers have tried so hard to erode![8]

The most important task of the guardian state will be to identify and enforce the massive emissions cuts that Australia must make as its share of the global effort to prevent runaway warming. If we work from the huge reductions scientists think we have to make quickly, our targets will be much higher, and will need to be achieved in much shorter timeframes, than the cuts presently being discussed within global or national frameworks. We'll need a state with powers to make this happen. Lovelock points out that warming might be slowed if we 'eat less and save energy', but knows that 'in practice we never will, unless made to do so'.[9]

Clive Hamilton observes, 'If we are to have a good chance of heading off the worst effects of global warming, emissions must be cut by at least 60% by 2030 globally, which means cuts of 90% in rich countries.'[10] To achieve this, 'all major democratic nations would need to elect governments wholly resolved to undertake structural change and override the most strenuous objections from the most powerful interests'.[11] Political debate and resolve in Australia today is nowhere near this. It exists in a dangerously separate parallel universe.

Hamilton indicates Australia's fair share of the global task of climate emergency response. This sets the targets for rationing at all governance levels. It would take another book, and one I am not qualified to write, to detail a global response. There are plenty of well-made accounts that do this.[12] The work of calculating national climate response allocations has already been done. EcoEquity, a US-based network of scientists and advocates, is working on these targets, which set high aims for the developed world, which bears most responsibility for climate change, and allow 'development space' (that is, some further growth) for less affluent nations.[13] Sadly, the community of national governments has not yet embraced this necessary and urgent work. When the inevitable emergencies force action at the global level, the turbo developers, China and India, are going to have to accept the fact that blindly emulating the Western development path is not a course of liberation. As Clive Hamilton says, 'It is a sad irony that the most powerful legacy of colonialism, the fetishisation of economic growth, may end up destroying us all.'[14]

We are also entering the climate crisis in the wake of failed expectations generated by neoliberalism. In the second global conflict, Allied governments suspended free capitalism on the promise of something better. This time, I think guardian states should be bold enough to promise something *different*, once the immediate crisis has been seen off. My broad speculations on what this difference might entail are outlined in the next part of this book.

How will a guardian state square with our sort of parliamentary democracy? The short answer is that politics and the state must function in something like the way they did during World War II. Party politics continues, but within a new bipartisan consensus that authorises strong central co-ordination for a time. Libertarians and communitarians will argue that power should be devolved to households, communities and individuals, who could find a Lilliputian way through the crisis through well-informed actions. This is a flimsy dream. Civil rights must be preserved by the guardian state, but this does not mean recourse to libertarianism. The period of emergency will require restraint of many habits and practices, largely to do with consumption and production. We might need to ban air conditioning, but measures like that wouldn't interfere with cultural and political expression.

One result of decades of neoliberalism and massive cultural change is that our social structure is not capable of carrying the burden of adjustment without a very strong co-ordinating hand. Structural reform has left us more dependent than ever on the state. Cultural and demographic shifts, many of them very welcome, have atomised our social structure, leaving an increasingly large number of us in small households. Another consequence is that many families do not have the support networks from local community or wider kin that were available in the last global war. As I showed in Chapter 4, a large and growing percentage of us now utterly depend on state payments. A central guiding hand is going to be necessary during the crisis, more so than in previous times of national emergency. We should not be happy with this, and must contemplate how to create a less 'statist' future after the crisis. Both Left and Right agree on this, but differ, of course, on what such a future would look like.

It's not my intention in these pages to draw a detailed diagram of the guardian state, but I want to specify a few premises and ends. This state will be needed, but it isn't an end state. Just as with the last global war, this will be a transitional state, with temporary powers to suspend, withhold and, if necessary, strongly suppress some urges, constituencies and structures (such as free markets). We must win through to something different, a society that does not need such centralised strong-handedness but which has a more decentralised resilience because it is generally more aligned to the imperatives of natural renewal. Are we clear? I am not urging a 'thousand year' strong state; I am urging its very opposite, a transitional state, just as in World War II, that knows it must dissolve at the first opportunity.

The contrast is what I fear, the 'eco-cratic state'. This would be governments built on a different interpretation of 'emergency rule', and which look more like our Axis enemies in World War II, or any other total state since. They would arise from the scenario I dread, in which we leave the politics of responding to our emergencies to the last possible moment. Consider a series of rapidly unfolding disasters (a city runs out of water, a storm lays a region to waste) that overwhelm our institutional capacities and cause widespread panic. We'd be exposed to the imposition of emergency rule 'in the interests of all' – and with 'business sector support' – that suspends civil rights and democratic processes generally. Indeed, fascism by any other name. This state might cloak itself in environmental reason – saving nature from itself, saving us from nature – but then set itself against the rights we now take for granted, especially civil liberty.

Increasingly, we hear the war scenario invoked by those who share my view that we face a species emergency. I think the analogy works in terms of the economy, and to some extent culture, because we will need to suspend and replace market circulation with new distribution processes that work back from nature's limits. Goods and services will be allocated and distributed centrally, the allocation perhaps based on minimum individual consumption rights, especially for necessities such as water and energy. Markets will still operate where they do not interfere with the goal of meeting strict quotas.

None of this requires the suspension of civil rights, as happened during world wars because we were under military threat. In the last war, the states of Queensland and Victoria had ominously named and empowered Committees for Public Safety. There were indiscriminate internments of enemy aliens without trial, even in those German and Italian communities that were avowedly anti-fascist. Under cover of war there were new brutalisations of indigenous Australians in the north of Australia, who, it was thought, couldn't be trusted in the event of an invasion. We mustn't repeat anything like these disgraceful episodes. This emergency, of a different kind, instead necessitates the suspension of ordinary consumption rights and the careful preservation of civil rights; civic values and solidarity must be strengthened, not undermined. The guardian state would safeguard the collective good by supervising consumption and production more closely, and by preserving civil liberties. It should also, in my view, begin to restore our most important and yet severely degraded democratic resource: the public sphere. In the wake of neoliberalism this state would do more to restore democracy than to restrain it.

There are two strategies for avoiding an autocratic scenario. First, we need to generate political change and response now, rather than waiting until things slide from bad to terrible. Remember that climate scientists are telling us that the tipping point, and thus the slide, will not be gradual; it may well be sudden, precipitous. The same applies to many of our natural resource stocks: they are threatened by collapse, not slow erosion. Geopolitical instability may snatch oil from us suddenly, for example. It will take a lot of effort to raise popular awareness of how close to nature's edge we are. All our production and consumption systems are pointed in the wrong direction, egged on by politics. The GFC stimulus responses of Western states from 2008 have aimed to shore up consumption, and have missed the opportunity to rethink it all from the production phase onwards. It's difficult politics, to be sure, because it implicates the market, as it is presently constituted, as an eco-cidal mechanism.

The second pre-emptive strategy is what I have been banging on about in the previous pages: reinstating equity, solidarity – and, while we're there, modesty – as central social values and guiding lights for government. We already know that fairer societies are happier and, critically, more resilient than unequal nations. The evidence for this claim, if you need it, is presented fulsomely in a recent book by British scholars Richard Wilkinson and Kate Pickett, *The Spirit Level*.[15] Its subtitle, *Why more equal societies almost always do better*, summarises the *factual* evidence about inequality that neoliberals ignored or scoffed at for years. The book presents an analysis of wellbeing in developed nations, including Australia. As the authors show, the social harm caused by inequality cannot be quarantined to the poor. Unequal societies generate morbidities, such as failing mental health, obesity and violence, that affect all social strata. Egalitarian societies, by contrast, have much lower rates of general dysfunction and are healthier, safer places for all citizens. In the stressful time ahead we will need the resilience gained through equality and transparent fairness more than ever. Maintaining, even strengthening, civil rights will be part of this. The guardian state must carry this ideal on its breastplate as we enter the survival battle we brought on ourselves.

Finally, let's get one other thing out of the way. To advocate human solidarity is not to push species chauvinism back on the planet in another form. All of what I've said above hinges on us letting go, completely, the Prometheanism that has held a bony-fingered hand on our hearts for centuries. This deadening claw must be seen for what it is, to invoke Malouf again: one of many failed experiments we must now release ourselves from. We are a species dependent on all others, and in turn, all life is dependent on the 'insentient' systems that comprise the material Earth. So our own commitment to solidarity is a starting point, not an end state, and has us committing outwards, to the preservation of Lovelock's Gaia. The guardian state must take this natural dependency on, not as a hair-shirt, but as a threshold for science, for policy and for governing.

And so it must go. I write these words as southeast Queensland, my home, begins to dry again after a brief respite from a prolonged drought that took us to the edge of order. Yet another record-breaking season has just passed: a winter that was warm enough to confuse the whole spring ensemble and rouse it too early from its winter torpor. August 2009 was the hottest such month on record for the eastern states. Thirty degrees days in winter Brisbane; does anyone even remark on this any more? The climate experts who tell us on the airwaves that this is 'consistent with long-run warming trends' are just so much white noise. Are we inured to climate change in subtle, whispered ways already? If such popular resignation already exists, consider how fabulously out of touch the climate denialists are. There is no army of common folk marching at their back. Barnaby Joyce is the piper leading a phantom army of children to the river caves of denial. The problem with the 'long-run warming' prognosis is that it neglects to point out that global heating will at some point produce sudden, cataclysmic change, not a continued gentle erosion of our winters.

The big change, when it comes, is likely to be a series of reverberating, interlocking shocks, perhaps triggered by a set of very 'extreme weather events', as the Americans might say. Who knows what political economic stresses will also apply an additional hand? I paint this scenario with a heavy heart because it seems we will continue to ignore the need for radical change until it is forced upon us. A friend once shared a personal wisdom with me: 'All my virtues have been forced upon me.' I think Australia, the West, perhaps the world, will thoughtlessly observe this maxim and delay change until it is forced down our throats.

10

LIFEBOAT CITIES

… we have a chance of surviving and even living well.
But for that to be possible we have to make our lifeboats
seaworthy now.

James Lovelock 2009[1]

THE LIFEBOATS THAT WILL CARRY *homo urbanis* to the next stage of its existence will be urban. This should be one further cause for hope if we wake in time to take the rudder and steer the best course possible through the storms of change we have brought down on our own silly heads. Why? Because cities are steerable (awkward word, I know) in a way that dispersed agrarian or regionalised communities may not be. Lots of us are assembled in city systems whose basic functioning can be manipulated. This is of course to step around the recent neoliberal tendency to allow cities to freewheel across natural and social limits. That 'model' (the urban vortex) must join the dustbin of history.

Our shift to the lifeboats may one day be seen as the saving of our species. But the lifeboats are far from seaworthy at the moment.

The urban shift has happened very quickly: a century ago, only 250 million people, 15 per cent of the world's (much smaller) population, lived in cities. More than one billion people now live in miserable, informal settings – massive squatter camps and barrios ringing developing cities – that are ageing, worsening and intractable to improvement. By the middle of this century, two-thirds of the world's population will live in cities; 19 of these cities will have more than 20 million inhabitants each.

The prominent US urban scholar Mike Davis, author of *Planet of Slums*,[2] argues that much contemporary urban growth is taking us 'back to Dickens'. Architects and aesthetes fear that the global economy is producing a world of suburbs: this is the anti-suburbanism I spoke of earlier. Davis shows that slums are proliferating much more quickly and with far more hideous consequences. Many of these festering settlements will be the first lines of defence, or defeat, as global warming makes sea levels rise. The prospect of nature in control of a game of slum clearance is too horrible to contemplate. But we must.

In Australia, we took to the boats early on. We have long been an urban, indeed suburban, nation. Our urban vessels have carried us very well through our short history, but we have also used them to run right over the original owners of this land. This was unjust and foolish. I believe that in turning these vessels into lifeboats we will have to turn to our indigenous brothers and sisters with humility in a quest for knowledge about country, and to learn to see the productive, nurturing Earth in new ways. This would be *real* enlightenment. We will need to gaze back with doubt upon all we hastily created, recognising bad and good within it all, and then seek new knowledge, which might include ancient wisdom in its energetic surviving forms.

Taking to the lifeboats means abandoning the ship of foolishness that we know as Prometheanism. We must think about how to manage ourselves in a long confined voyage to uncertain shores. None of this is to abandon rural and regional Australia, but its fortunes will largely follow in the wake of our cities. Each urban population will face that holy trinity of survival values – restraint, sacrifice, solidarity – in somewhat different ways, conditioned by its unique history, ecology and society. A general challenge will be to define and practise what Lovelock describes as '[t]he ethics of a lifeboat world'. These are 'wholly different from those of the cosy self-indulgence of the latter part of the twentieth-century'.[3]

Just as in the lifeboat tales we know, the basics will be finite and will have to be rationed amongst stressed passengers, some of whom will be restive. Basic tasks, including the bailing out of an encroaching nature, will have to be shared out. Everyone will need to surrender much personal space and all preciousness in what will feel like an ever more cramped situation. But there will be prizes along the way: solidarity is its own social reward, and our ability to face adversity should see us 'cop it sweet' with good humour. Let's hope we haven't lost that last facility – there was an awful lot of self-preening and whingeing during the neoliberal era. Aspirationalism exposed a newly narcissistic side to Australia that will not be a source of resilience in times of trouble. Baby boomers especially, take note.

An 'urban system' is a set of interlocking systems that feed, house, clothe, employ, educate and generally supply our needs and wants. These production, circulation and consumption systems have produced fireworks during free-wheeling neoliberalism, but are not working to sustain the urban whole. The evidence was presented in Part 1: a shelter crisis for many, ever-lengthening commutes for most, and scarcity in key natural resources, especially water. Even if we were not entering an emergency, the continued functioning of our cities would require us to repair these 'wild mechanics' of neoliberalism. This means clearer, stronger metropolitan governance for our cities; they cannot any longer be left to the vagaries of the market or the malign neglect of some state administrations.

I believe we need to move immediately to establish strong, independent metropolitan commissions to lead, not simply manage, our cities through the coming crises – to steer the lifeboats. They must have statutory independence and democratic input, including representation from the municipal layer that they will co-ordinate. We've been there before. From 1891, the Melbourne Metropolitan Board of Works provided a very successful model of this co-ordination and co-operation; it was not dismantled until the chronic 'reform' era of the 1990s. The metropolitan commissions will be parts of the wider guardian state apparatus, working in concert with state and federal governments to recraft our systems of resource allocation.

In the search for sustainability and security, the Commonwealth must broadly supervise and support the urban lifeboats that will carry most Australians through the crisis. This will require a multilateral effort to establish cross-party support for federal urban policy. During his long prime ministerial tenure, John Howard was adamant that planning and urban governance were state responsibilities and no concern of the Commonwealth, much to the vexation of urbanists – and, interestingly, the business lobby. With his exit from the national stage, I sense that conservative politics is likely to transcend this foolishness. There exists at last the possibility for a political consensus on the need for national urban policy of some sort.

A national settlement strategy must be devised and implemented in partnership with the states, territories, city commissions and local governments. It should be one part of an 'Australia Plan' which will include economic, environmental and social strategies for bringing us through the climate and resource crises. We've dragged our feet on the idea of a population strategy for some years now. This cannot continue. A population strategy we must have – it will be impossible to recraft our settlement pattern without it. There will be difficult decisions to make about internal migration and immigration as our brown land dries out and sea levels rise. As part of this we'll have to decide which parts of our land we'll have to depopulate – probably poorly watered rural areas and some grossly exposed coastal communities. This discussion has already begun in some frontline coastal areas, such as Byron Bay in New South Wales, where it's obvious that some exposed urban areas will not be defensible. But these difficult decisions cannot be left to local governments who don't have the resources to manage the inevitable planning disputes and compensation claims that will arise as retreat lines are marked out.

The Australia Plan will focus on making Australia secure and sustainable, and on the maximisation of wellbeing, not material wealth. This national settlement strategy should be grounded in a thorough research-based assessment of Australia's settlement pattern, examining the sustainability and security threats that face our cities and towns and the best ways in which these might be swiftly countered. The climate change emergency must take high priority.

Each city commission will use a set of resource targets set by the Commonwealth to guide their work of rationing and distribution. There will be room for some flexibility based on local capacity, especially in terms of generating water and food. The most fundamental allocation frame must be energy use and CO_2 emissions. These allocations must be co-ordinated at the national level, and then flow through to the states and city (and presumably regional) commissions.

The city commissions must nourish, not suppress, democracy as far as is possible in a time requiring firm central co-ordination. They will have powers to circumscribe consumption rights but not civil rights. And the commissions will be charged with maintaining a flourishing civil society as a means for strengthening solidarity and for accessing 'non-expert' knowledge. What does the latter mean? It refers to the pool of usually unspoken and unremarked wisdom that resides in the community and that rarely comes to the attention of governments at any level. There are already legions of sustainability innovators in our cities, especially in our vast suburban landscapes, from whom we have much to learn: that is to say, we can even now *learn from the suburban Earth.* A bright example is the international Transition Towns network, which has a growing presence in Australia.[4] This is civil society at its best – networks of citizens generating and testing new means for repairing or overcoming our failing urban resource systems. This is not merely technological innovation; it is democratic strengthening, producing new conversations and alliances that sometimes surprise all involved. The city commissions must respect, nourish and mobilise such civil resources.

There looms the question of getting things about – humans, goods and services. The Commonwealth will need to drive a complete renovation of our transport policies and support the metropolitan commissions in this work. Australia must discard transport policies that massively favour private motorised travel over public transport and non-motorised forms of transport such as walking and cycling. We've heard plenty about this for years but done next to nothing. Now, in the face of interlocking greenhouse and resource crises, we will have to rethink our transport systems just to maintain the safe functioning of our cities. The emphasis on individual motorised mobility must be replaced with the goal of maximising urban accessibility for all citizens. This will require a great rebalancing of resources and strategic commitment away from private motor travel. As part of this, the massive recent expansion of air travel will also need to be reined in.

The national broadband system, presently under construction, must be a truly civic resource – accessible by all, irrespective of wealth – that will provide information and services directly to people in a time when travel will be expensive and curtailed.

The city commissions must rewire our urban systems around a central processing logic which will do two main things. First, they must ration the consumption of scarce necessities (such as water) and harmful things (such as non-renewable energy). My belief is that rationing should accommodate fundamental resource use rights, which guarantee – let's hope it remains possible – a minimum amount of resources (especially water, food and energy) to each urban citizen. My preference is that the guardian state provide a basic per capita allocation of water and energy for free, and heavily price whatever discretionary consumption is still possible within rationing limits. This would give everyone a basic sense of security.

It will surely be necessary to restrict, perhaps even ban, the unlimited use of harmful technologies such as air conditioning; it must be reserved for special places, such as buildings for vulnerable people – the aged, the sick and children. This won't be easy in a time of rising heat stress, and the commissions will have to help urban populations achieve climate safety through low-carbon practices sincluding minimising travel and fortifying homes and workplaces against weather extremes. This would mean a return to the ways our parents and grandparents lived before there was universal access to cheap electricity and appliances that regulated building climates. And many innovations, including better home insulation and shading, will guide us towards the low-carbon lifestyles we must live.

A lot of energy will have to go into designing workable distribution systems based on rationing, and into extracting more, much more, from what we have. Households must be helped to produce as much of their own food, and capture and generate as much of their own water and energy as possible. Again, fairness and solidarity must reign: some people will need and deserve more help than others. Suburban people with big yards may need to join a larger effort to sustain cities by sharing some of the produce they generate. Wealthier people in highrise communities could volunteer labour to support suburban food production in return for a share of produce.

The second major task of the commissions will be to improve urban system functioning. This will necessitate the better alignment of housing and labour markets; that is, people should be living closer to where they work. This will require more directive planning. A large share of occupations should be grouped in district centres within our extensive cities. These centres will be public transport hubs, requiring an extensive investment in well-integrated trains, light rail and bus services. We have to get over the misleading idea that suburbia is anathema to public transport. Cities in other parts of the developed world, such as Toronto, provide high-quality and well-used public transport in suburban regions. The case is presented clearly in Paul Mees' book, *A Very Public Solution: Transport in the dispersed city* (2000).[5] Public transport will emerge as the fundamental lifeline in suburbia when household vehicle use is suddenly curtailed by oil shortages, and energy rationing to meet CO_2 mitigation targets is steadily enforced.

The dispersal and reconcentration of activities, opportunities and distribution networks within metro district centres will help make suburban populations safe during a time of stress and dislocation. To meet the tests of efficiency and fairness we will need a strong circuitry within our cities that places nodal points of opportunity and supply as close as possible to our suburban communities. Our current metro strategies identify such centres but do little to shift employment and activity towards them.

Just before you decide that my vision of urban guardian states – metropolitan commissions – is a necessary (or not) democratic brake, consider the current governance of our cities. If we start from what we have now, there's not a lot of democracy to lose or trade away. Let's examine one dimension of this. A huge 'tollway industrial complex' of financiers, consultants and construction interests has been turning our cities inside out for the past decade or more at massive expense, and usually in the face of community concern.

Consider a recent national infrastructure conference reported in *The Australian*. It was presented with survey evidence which showed that there was no general public support for Public Private Partnerships, the finance and governance model that has placed an increasing amount of our urban infrastructure in private hands. This conference conversation between finance and industry was advised by one expert delegate that 'the aim should be to convince government and politicians that political risks could be managed rather than trying to win over the community'.[6] Sometimes you don't need to satirise; you just let the people speak for themselves, or at least make sure their secret confessions are heard.

This reveals how contemptuous of liberal (let alone social) democracy parts of the urban infrastructure lobby have become. Returning fundamental urban decision making to a democratically accountable body, a metropolitan commission, would be a major advance on what we have now. Again, this is an improvement that has to be made even in the absence of the current emergency. More generally, the guardian state offers the prospect of taking back democratic control of many of the basic steering mechanisms, including infrastructure planning and provision, which were partially surrendered to the corporates under neoliberalism.

Part of this will be halting the privatisation of existing and future infrastructure, especially the services that provide our critical resources: gas, water and electricity. Resource conservation fundamentally contradicts the motivation for private sector involvement in resource infrastructure. It's another recipe for overproduction. Environmental scholar Sharon Beder writes:

> [I]n 2006 stationary energy sources generated 40 percent more emissions than in 1990. This massive increase has been aided by electricity privatisation and deregulation in Australia which has provided little incentive for electricity companies to promote energy efficiency or renewable electricity.[7]

She further explains that 'deregulation and privatisation have led to the increased use of the most polluting type of coal, brown coal'.[8]

So subsidise the recalcitrants to innovate and conserve? Emissions trading schemes and dedicated subsidies are bad value for public money and risk transferring more wealth to corporations. It will be more effective and cheaper to resume community ownership and steering of these assets and resources. This is not to encourage the preservation of monoliths. After the crisis we must decentralise and democratise our infrastructure systems in order to ensure long-term sustainability and security. The reliance on big network infrastructure to provide urban services must be discarded in favour of a more decentralised system where householders and firms assume more responsibility for on-site water collection and treatment, power generation and food production. This must involve public investments and subsidies to help households and firms achieve water, energy and waste independence and to encourage the development and take-up of renewable energy. Urban infrastructure could return to localised – municipal or communal – management, guided and supported by policy and funding from the metropolitan commissions.

We must also think about whether rollbacks of privatisation will have to occur, for example in states that have sold off parts of their water and energy systems, such as Victoria, South Australia, Queensland and New South Wales. As Beder points out, 'Only governments, which don't require high short-term returns on their investments, can make the concerted effort to invest in the sort of technologies necessary to prevent further global warming.'9 Agreed, but even before we get to the question of system adaptation and improvement we will need states to allocate and co-ordinate our use of energy, which must be heavily rationed for a time. Eco-cidal technologies will have to be dispensed with, including desalination powered by non-renewable sources. Some of this will make the rationing and allocation tasks harder, but we simply must meet cutback targets.

We already have a shelter problem, which will only get worse as economic and social dislocations arise during the emergency. The metropolitan commissions will have to lead the creation of, if not completely deliver, new 'no-nonsense' urban communities. They will not be based the contemporary super-frills masterplanned estate but on fundamental needs: room for all, solid social facilities and amenities, walkability and good public transport. The housing can be straightforward and adaptable, but of good quality. It should be a mix of tenures, including housing provided by community and public agencies. Public and social housing should be dispersed throughout our cities and towns.

Both socially supported and working people should be offered further opportunities for housing in an expanded social housing sector based around independent, publicly subsidised associations, some with specialist missions, such as supporting the frail elderly and supporting re-entry to community for homeless people. This will provide humane and inclusive alternatives to the warehousing of aged Australians in corporate aged care facilities or gated retirement communities.

Access to safe, affordable and pleasant housing should be guaranteed through new tenurial and finance options that prioritise security over ownership. Tenure security will need to be strengthened, meaning stronger rights of guaranteed occupation for people in private rental housing. Landlords will have less speculative freedom but more certainty through contracts that will insist on good maintenance and conduct in return for longer, protected leases.

The current reinvestment by the Commonwealth in the urban public domain is beginning the task of building social resilience. Government economic stimulus programs are targeting our long-neglected public realm: our decaying hospitals, civic amenities and schools are all receiving new investment. Federal funding to renew ageing public housing stock and to provide new accommodation for the homeless will bring back into circulation many units lost to disrepair and provide a long-overdue facelift for many. It marks a major turnaround. This work must be carried much further by the metropolitan commissions, with national financing.

Life in a time of stress will not be bleak if we value the many good things that will flow from a society refocusing on modesty, solidarity and fairness. Historian Raymond Evans recalls the 'terrible novelty of total war' during the last global conflict, when all sorts of established cultural values and social mores were challenged, sometimes tipped on their head.[10] One important one was the liberation of many women from stultifying sexual identities and social roles. I suggest we think positively of the wondrous novelty that might be possible as we enter a time of total conflict. Emancipation from the soulless individualism of neoliberalism is on offer.

James Lovelock seems almost gleeful. Having lived through the last world war, he is less apprehensive about what is to come than I am:

> Just as in 1939 we had to give up on a massive scale the comfortable lifestyle of peacetime, so soon we may feel rich with only a quarter of what we consume now ... For the young, life will be full of opportunities to serve, to create, and they will have a purpose for living ... Whatever happens, it will be quite a change from the banalities of city life now.[11]

As our overcooked cities start to cool, we might get time to rethink things, with children foremost in our minds. The time of stress will involve economic dislocations that profoundly recast and upset working life for many. Unemployment is a tragedy, but underemployment will be equally challenging. Many presently in the workforce will be forced to work reduced hours. The project of restoring urban food production could use this surplus labour power, with participants recompensed with food or other 'exchange credits'. For many of the over-workers, and especially their families, this will be a blessing. We might start taking seriously the goal of 'work-life balance'. We're seriously imbalanced now, working some of the longest average hours in the world. A restoration of family time will be one bright light in a time of gloom for many.

And restoration of personal time also beckons. Urban communities are starting to refocus on their neighbourhoods as places to work and play. Backyard gardens and community plots are flourishing, and the recession will drive their continued growth. As traffic is calmed, community-focused cities will be safer and more stimulating for kids. Urban civil society should be restored as part of a general renewal of social solidarity and purpose. The tendency of governments to favour private provision of urban services and management of urban spaces will be replaced by a new approach founded in confidence about public endeavour. The city commissions will ban gated suburban communities. Exciting, safe and well-supported public spaces will be established in these suburban areas of our cities, managed by community or local authorities.

We enter the coming storm as yet ill prepared, our urban lifeboats far from secured and provisioned for the tough journey ahead. My speculations above are a very partial plan for what must be a comprehensive effort to take Australia to safety, or at least a place of survival, as climate and resource systems start to work against us. What we have lacked to now is leadership to steer us from the course of waste towards a society that values stewardship, renewal and fairness. There is evidence, however, that Australian civil society is demanding that its leaders reset the compass of politics away from the short-term thinking that has left us vulnerable to social and ecological breakdown. The new setting, towards sustainability and security, will require a changed urban course that will take our cities through the storm ahead to more resilient shores. Like it or not, our journey to the next world has begun. Let's hope we can have some influence over what it looks like. The question is, what do we want it to look like?

11

LINING UP FOR CHANGE

SO WHAT MUST WE DO to make things work during a period of shortage and stress? To end this part of the book in a lighter key, I offer now a story of change which starts with our current absorption in consumption. If a crisis of supply should hit us, as natural systems collapse or suddenly pull their boundaries in, how should we respond?

My answer is directed towards the guardian state, which must move swiftly to secure our material lifelines, once Promethean neoliberalism has been called off – probably, finally, by a series of resource emergencies sourced in climate shifts and in an ever touchy global economy. We will be forced to respond to these emergencies and the privations they'll bring by pulling our heads in to meet severe emissions reductions targets. We're running out of time to avoid the point where global heating becomes runaway climate change. The longer we leave major action, the harder we will have to be on ourselves in reining in production and consumption.

What might all this signal for us happily(?) consuming Vegemites? It will force a big change to our consumption and distribution systems so that, to repeat Paul Kingsworth, even 'ethical shopping' will have to give way to different ways of getting what we need. The news is that we are going to have to embrace the queue again. The same enterprise that went into trying to make the queue extinct has made it necessary again. The growth machine economy of the past few decades has gobbled up and despoiled much of what we need and desire and shortages are already emerging.

Water, petrol, energy, green space and food are all in increasingly short supply, and things will continue to tighten. We are out of time for techno fixes or market adjustments. Many things that are in short supply and/or cause environmental harm are going to have to be rationed. We'll have to find ways of sharing scarce, damaged and damaging resources and stop imagining that they are immune to injury or exist in endless abundance.

Like you, I hate queuing. Partly because I'm bad at it. At the ATM I always seem to be behind the person who needs to type their biography into the unco-operative machine. In the supermarket I always choose the checkout that's landmined with price checks. At the video store I end up behind the bloke with hearing problems trying to take out membership. To avoid this I visit different video stores, but he seems to be trying to join them all.

There are a few boobies and saints who don't mind queuing. But not you, not me, and not just about anyone else. In the past, however, we were better about it and better at it. Think of how many of those old black and white photos we see in pictorial history books show patient, functioning queues … for buses, for soup, for autographs, for a glimpse, for a thrashing. How many of us associate queuing with privation ('Just one cabbage each, and don't push') or punishment ('Your turn, son, six of the best, put your hand out …')?

Queuing is a social experience, a collective act. We generally don't get to choose who we line up with. And the line order usually ignores the social order. In real queues money or connection can't get you a better place. A great democratic sight is the unhurried pensioner at the postal counter, holding up a row of highly strung suits. But like dogs with buckets on their heads or old ladies driving with fags in their mouths, it's a rare urban sight these days. The queue's a scruffy relic in an age of universal individualism, momentary gratification and compulsory choice. Much of our technological innovation and economic reform has been aimed at annihilating the queue. Some of this has improved our lives, but some of it hasn't. The infernal need to 'choose one of the following 9 options', and the threat of being recorded 'for training purposes', has me thinking fondly of the post office … sorry, post shop.

The queue is a wonderfully simple and effective rationing device. It's fair, and deaf to complaint. Unlike the 'customer service representative', you can't argue with it (mind you, that type of argument is pointless). We haven't completely lost our queuing know-how: witness the masses surging back onto urban public transport systems as petrol prices soar. We've handled it pretty well; there've been no reported murders. We have enough of the queue's DNA to rescue it and clone it so that it can repopulate our imperilled cities and communities. You will soon see all sorts of queues popping up around the place. And you will see lots of them first hand, as you take your place in a line of people waiting for something.

The first bloom of the queue species will be for things we want but don't need: a spot in an overused place such as a national park or a camping ground, or for grossly environmentally destructive activities, which may ultimately include flying. You may scoff, but we once rationed luxury goods such as imported sports cars, though for other reasons. Rationing luxuries such as resort stays isn't pretty. We're unlikely to go the full Soviet holiday-camp-for-all route, for sensible reasons, so rationing will put this form of escape beyond the reach of many who now enjoy it. It's likely that increasingly scarce goodies will be captured by the rich and the well connected, as they have always been.

In the long run, and as the screws tighten, there will have to be fairness in the distribution of scarce things, or there will be trouble. We may have to generalise some of the discoveries of the market, such as resort time-sharing, to manage the distribution of increasing burdens and declining opportunities equitably. Transparently fair adjustments will help us maintain the social solidarity we will need.

Rationing doesn't mean we have to make people line up for everything. Households can be allocated fundamentals according to need. Pooled provision and use will also advance the cause of rationing. The plasma screen will not be ubiquitous and we'll have to learn to go to the cinema again. The queue will be an important weapon in the war on waste because it is a powerful restraining device. If there's a queue between you and what you desire, you have to be pretty sure you want it, because there will be work and time involved to get it.

This is not a misty-eyed argument for Soviet-style collectivism. It's not a socialist utopia that's being urged on us; it is a situation we brought on ourselves. Much of the rationing experience will be painful and confusing. There will be squabbles, but, let's hope, not worse. Some of the things we enjoy now simply won't be possible and we'll feel the poorer for it. Things like a recent internal European flight that cost me £6. I took this flight on a budget carrier to deliver a public seminar on climate change. I'm part of the problem. The pain of change will be felt personally. I'm married to a European. I feel immensely worried that the end of budget air travel will shrink her world and push her family and homeland away. This isn't going to be easy and lots of people are going to feel disappointed, stressed and cross. The guardian state will have its work cut out just to deal with the tsunami of disappointment that is sure to follow the collapse of neoliberalism.

However, there are riches on offer if we know how to identify and value them. Returning to a society of modesty, solidarity and locality might not be good for Harvey Norman but it could enrich us. Rationing will slow us down and we might see more, not less, of each other. Importantly, parents might spend more time with their kids and take a more active part in their lives. This could be the most effective way of stemming the processes that are making them fatter, sicker and sadder. Traffic-calmed cities will be cleaner and safer; again, especially for children. If we walked more, we'd be thinner, healthier and happier.

It may not be *The Da Vinci Code*, but the mystery of the queue may be that it is likely to eventually quiet our increasingly agitated hearts and minds. We might better understand that our whole lives are a sort of queue over which we have no control. All of us will leave the Earth in a sequence we cannot know. We might more calmly ponder and accept that one day, like it or not, our number will be called. Animals don't queue. We invented the queue because we know its power to protect us from self-destructive habits. This time the queue might save our species. Worth thinking about next time you're in the supermarket deli behind the person slowly ordering a bit of everything.

TO THE NEXT WORLD

WELCOME TO THE NEXT WORLD and the new Australia. It's hotter, drier and more tempestuous than the world we left behind in the 20th century. In the great transition that followed, many grievous things happened. Forced migrations, wars and climate catastrophes cut a swathe through many human populations and many others suffered times of great privation, uncertainty and anxiety. Nature became aroused and inflamed and turned on itself; there was a spectacular collapse in biodiversity in many regions, and most lost beautiful things and places. The sphere of habitable Earth receded. Many formerly productive areas in Australia turned to dust, coastal areas crumbled away, and the city lifeboats heaved their ways through turbulent seas …

And yet the worst did not come to pass. James Lovelock's vision of a human species dwindled to small breeding communities in just a few oasis regions was not realised. In the face of many disasters and brutally compelling evidence, the world community of nations was finally moved to act and co-operated to enforce massive carbon emission cuts and the securing and sharing of vital resources. Australia's guardian state brought us through the worst, with most passengers accounted for. The lifeboat cities witnessed much stress and not a few petty acts of mischief and rebellion, but they made their way to a more settled, if challenging, biosphere. Carbon emissions were radically curtailed and the climate began to stabilise. The new weather regime was gradually better understood and we learned to live with it. We developed many new adaptive techniques, and they began to restore productivity and provide basic resource security. Resources such as water, food and renewable energy became cherished, but the sense of emergency receded. There was limited resettlement of some of the devastated and all-but-abandoned regional areas. We learned how to restore their productivity within the confines of new climate realities. The world's oceans experienced a slow turnaround; acidification gradually receded and some fish stocks re-emerged. Many things were lost forever, but some things thought lost were regained.

And some wondrous things happened. The foul miasma of scepticism and denial was swept aside. A consensus for change and survival emerged and a bright sense of changed social purpose glowed. Respect for nature at last started to flow deeply through our veins again, and the platitudes of the 'green' past were left behind. Social solidarity survived many tests and flourished. Long-held assumptions and ideologies fell to the wayside. The declaration of an 'end of history' made in the late 20th century by neoliberals is looked back on as a treacherous prank. It diverted attention from the desperate need for a new political and social dispensation ... and for a new compact with nature.

Slowly emerging through the clearing mists of struggle was a new idea of human flourishing. Finally our species developed a willingness to avoid the cycle of growth and collapse that has directed our course since the first times. It took a crisis that nearly destroyed humanity to teach us the meaning of critical self-awareness – our most potent yet heretofore most latent power. As the crisis lifted, a general willingness to change was evident. Time was called on the guardian state. A new social compact was determined: it was to guide the human species to a more secure relationship with the Earth.

And there was an even heavier silver lining in the clouds that accompanied us through the storms. After an age of catastrophic indecision and indolence, we recommitted ourselves to the project of modernisation that had long ago brought us out of the twilight of primitivism. We accepted that the project had not been fully realised through industrial capitalism. And that the purgatory of human ambition was not a safe place. We knew that the half-realised dream had turned into a nightmare of self-hatred. The bloodied improvidence of the 20th century was testament to that awful truth. Humanity in flight from that reality would no longer seek consolation in Prometheanism, authoritarianism, and fundamentalism.

We resolved to right our wrongs against ourselves and nature by refounding the great project of human liberation we knew as modernity. We dedicated the next world we had inherited to the realisation of human potential, taking with us the values that had seen us through the emergency: solidarity, modesty and justice. It was time to sweep the queues and the ration books away, to end the carefully prescribed behaviour of the hard times, and to realise our species potential for spontaneity, love and renewal. This wasn't a journey to utopia, but a firm resolution to reject the road to its antithesis, dystopia. We'd strayed there for a while and learned our lesson. *Homo urbanis* sought a new home, in the good city. This was the place for a more wakeful life. The fever of ambition and the dream of supremacy faded, and the work of realisation began.

12

TO LOVE FREEDOM

It is not for us, the staid lovers calmed by the possession of a conquered liberty, to condemn without appeal the fierceness of thwarted desire.

Joseph Conrad, *Under Western Eyes*, 1911

TO LOVE FREEDOM IS TO REGARD IT as a real objective, not a fantasy of species ambition. It also requires us to recognise that we are a long way yet from realising freedom. Indeed the difference between the immeasurable praise heaped on this ideal and the extent of its accomplishment has never been greater than now. Just how great is this deficit and how might we seek to close it?

We emerge from the cruellest, most destructive century our species has ever inflicted upon itself, propelled by vanity, greed and a good quotient of stupidity into an uncertain world, where our survival is hardly guaranteed. We confront a series of terrifying natural and social crises, which I have linked to the overproduction of the market. But that, of course, is only part of the story. We must also acknowledge the destructive legacy of our stunted social development, a type of development which was never acknowledged or addressed by industrial capitalism or its 20th century alternatives and foes.

The German philosopher Erich Fromm has much to tell us about this problem of stalled development in his 1942 book, *The Fear of Freedom*. Fromm left Germany for Switzerland after Hitler gained power in the early 1930s. Not long afterwards he moved to the United States, as did many German intellectuals, where he was able to contemplate two great convulsive models of capitalism: the fascism into which Germany had descended and the consumerism that was engulfing America. Nazi Germany was revolting against the modern ideal of human liberation. That much was clear. But what was less evident to the broader public was the manner in which early consumer capitalism, especially in the United States, was departing from the cause of freedom. Fromm and other displaced European intellectuals, especially those associated with the German 'Frankfurt School' of thought, brought this inconvenient truth out in their writings in the build-up to World War II and afterwards.

The Fear of Freedom was born of Fromm's observations in the 1930s and published in 1942. It carries this twin message of modern disappointment. The demise of feudalism was not an imposed destruction of a cherished way of life. We smashed our way out of a society that held us down with 'primary bonds': human servitude and our vulnerability to nature. We were eager to throw aside vassalage and happy to put great distance between ourselves and nature, which so often failed us. The stultifying culture of the village was rejected for the anonymity of the town, where freedom was protected and ambition assumed. The emergence of capitalism and the 'individuation' it spawned was a project of species liberation. A certain measure of freedom, of individual autonomy, was won for most, if not all, by the rise of market society. The great human breakout that we know as 'modernisation' was guided by a grand philosophical awakening, the Enlightenment. It remains a compelling vision. Most of the rest of our species still eagerly embraces the modern outlook, though its clumsy, often brutal, imposition through globalisation has generated many rebellions, including the revolts of fundamentalism (both Muslim and Christian).

Modernisation is hardly 'modern' in the ordinary sense of the world. It has been a long and often bloodied march. Centuries later, there is no turning back: our bonds to our community and to nature, once broken, cannot be restored, unless a total failure of modern civilisation occurs. The latter scenario has been embraced, even rejoiced at, in the apocalyptic literature that emerged during the Cold War, when we appeared to be heading towards species suicide through nuclear proliferation. The crises which I have focused on in this book – climate and resource failure – have given new vigour to the catastrophists of science and literature. Cormac McCarthy's 2006 post-apocalyptic novel *The Road* has been described as the first in a sub-genre inspired by global warming. Sections of the green movement, too, would have us unwind modernity now to avoid its inevitable downfall.

Apocalyptic visions may help to shock sedated publics and rouse some energy for crisis responses, but it is not my premise or approach. If it were, there would be no need for the guardian state and the whole project of survival and renewal I outlined in the previous section. What would be the point of learning to see the Earth if apocalypse is all we can expect? Better then that we learn to smell the Earth in which our civilisation would be interred. No, I believe we will rouse in time, and deploy our best values and technologies to bring us through the crises. But not unscathed and unchanged. However, for some, merely to speak of crises, of deep and destructive malfunctioning in our economic systems, is enough to raise the charge of catastrophism.

I am certain that we will come through the crises we face with our deep predilection for modernisation intact. A mass return to nature, somehow in expiation for our industrial sins, will not take hold in our hearts. The green movement, or sections of it, would be well served by ditching this errant hope now. The time of peril and pain we are entering offers not a negation of modernisation, but awareness that its promise remains only half-born, partly realised. I think it best to see the project of modernisation as the progressive emancipation of the human individual from those bonds that stymied our enormous, unrealised capacities for fulfilment. Originally these bonds were natural, so we had to achieve a certain degree of separation from and control over nature to progress. Later, we forged new bonds for ourselves as the liberation project stalled – authoritarianism, market consumerism and exploitation of nature. In 1911, Joseph Conrad, a great modernist, placed a rebuke in his novel *Under Western Eyes*. We Westerners were cast as 'staid lovers', anaesthetised by the small measure of 'conquered liberty' we had gained, and too often condemning 'the fierceness of thwarted desire', the yearning for a higher level of human fulfilment and liberty which surely inspired the Long March from feudalism.

Fromm too writes about this with poignancy, observing in 1942 the mass flight to fascism that propelled Europe and then the world to war and a new act of industrial savagery, the Holocaust. His view was that industrial capitalism was just one step towards emancipation. Fromm's argument is worth citing at a little length:

> [M]odern man, free from the bonds of pre-individualistic
> society, which simultaneously gave him security and limited
> him, has not gained freedom in the positive sense of realization
> of his individual self; that is, the expression of his intellectual,
> emotional and sensuous potentialities. Freedom, though it has
> brought him independence and rationality, has made him
> isolated and, thereby, anxious and powerless. This isolation is
> unbearable and the alternatives he is confronted with are either
> to escape from the burden of this freedom into new
> dependencies and submission, or to advance the full realization
> of positive freedom which is based upon the uniqueness and
> individuality of man.[1]

The argument is that capitalism and modernisation had by the mid 20th century produced an expectation of individual freedom, but had overlaid that hope with structures of conformity, especially in the spheres of work and consumption. In paid work life, most were wage labourers with little real autonomy and, just as importantly, no real power over or meaningful connection to the objects of their labour – commodities for sale and profit by others. In the sphere of consumption, there was individualism in name only. Indeed, the act of consumption was not a unique expression of need, but a moment in the ever-pulsing demand for commodities that were crafted to arouse, not respond to, desire.

Regularly, the covers on these superficial forms of individualism were blown away by system failings, including the economic Depression that followed the boisterous 1920s. Booms may blanket our senses with the white noise of avantgardism and hedonism, but they do not, as we have seen more recently with neoliberalism, give us a more secure footing in the world. Many in Germany and other European nations observed the frivolity and decadence of the 1920s with mounting distaste. Add to this the continuing pain and grief caused by World War I, and Germanic Europe was set for the mass expression of disenchantment. The flight to authoritarianism was a rejection of half-baked individuality. We all know what this meant for world history: modernity had to fight for its very life against the tide of Nazism. Another form of authoritarianism, the Soviet Union, also played a decisive role, bludgeoning the fascist advance backwards. Its own dreadful qualities were then exposed.

The restoration of peace and the construction of a new capitalist order after World War II did nothing to address the problem of half-born freedom. The Allied victory resecured the modernist project and suppressed its rogue forms (Nazism, and eventually state socialism). From there, however, things began to slide again. By the 1960s a range of critical observers were pointing to the moral vacuousness of the postwar boom. Much of this critique was anticipated by Fromm in *The Fear of Freedom* and in his 1955 book, *The Sane Society*. The new consumerism of the postwar revival and the demographic (baby) boom was described as descent to an infantilised, 'robotic' society. As early as 1942 Fromm was warning that our sense of freedom was being manipulated and constrained by the rise of mass consumption and marketing, to the point where it was meaningless.

For Fromm's compatriot, Herbert Marcuse, the rise and rise of consumerism meant that individual autonomy was all but a fantasy for the *One-Dimensional Man* (1964) of Western society.[2] Consumerism meant new heights (or depths) of conformism and social control by the systems of production, marketing and sale. This was not the loss of freedom but its perversion, with consumers manipulated to the point where we acted against our own interests, buying to fill implanted wants, not needs, and with no thought about the problems of waste, environmental despoliation and long-term renewal. Some of the critiques got a little offbeat, producing conspiracy theories about 'agents of control' somewhere in the cockpit of the system. They missed the point: markets, not evil agents, were now liberated (though often with continuing state support, even subsidy) to achieve their purpose – permanent expansion.

This was a pretty grim view of human possibility. Inevitably, rebellions boiled forth: the civil rights movements, hippies and Aquarianism, the violent counter-surge of the new left in 1968 – all were united by a desire to seek release from the burden of Fromm's 'negative freedom' and 'robotism'. The global economic crisis of the 1970s and the ideological and political triumph of neoliberalism in its wake put an end to the fun. It raised a new flag of liberation: the 'economic man', the 'self-maximising' creature of neoclassical theory. The rest, as we know, is recent history. I've pointed to some of the contradictions of the neoliberal project, such as the recasting state service users as 'consumers', which did nothing to improve their liberty and autonomy, but instead helped justify and effect a wholesale privatisation of civic effort.

And for the rest of us, I am certain the record of neoliberalism shows a diminution not enhancement of individual freedom, despite the rhetoric of 'choice' that thrived during the Howard government's reign. Social polarisation and impoverishment for some meant a loss of economic liberty. For the employed, contracting out, subcontracting, 'liberalised' employment provisions, longer working hours and constant, indeed chronic, restructuring produced a growing sense of insecurity and stress. This was powerfully documented in Michael Pusey's charting of quiet despondency in the middle class.[3] In the final years of the last century, this darkening of mood and prospect was ignored by the politics that celebrated suburban 'aspirationalism'. Suburban improvers were cheered over the personal debt cliff, encouraged to consume as much housing and domestic goods as credit cards and housing finance would permit. The GFC and spiralling resource (especially oil) costs have proven the folly of this course. The result: teetering suburbs in cities that are themselves on the edge of possibility.

And yet the apostles of aspirationalism continue their evangelism. To criticise this crumbling deity can still arouse a fierce defence. In September 2009 Mark Latham, point man for aspirationalism during his term in public life, charged me with 'random thoughts and hyperbole' for criticising this model.[4] I was deluded 'by a leftist expectation that the masses are about to snap out of the false consciousness of materialism'. My 'macabre warning' that Australia was waking from a dream of denial – the very subject of this book – offered a 'nightmare' vision as alternative. Quite what my alternative was in Mr Latham's mind was not clear. But he was in no doubt about his preferred suburban outlook: business as usual (economic and resource crises notwithstanding) with a bit more consumer choice in public facilities. Mr Latham casts my social vision as a hell. His vision is not a hell, but a purgatory of endlessly deferred human fulfilment.

He seems to have lost the bitter doubts he expressed about aspirationalism in the confessional introductory essay to his 2006 *The Latham Diaries*. The subsection entitled 'The Sick Society' reveals his personal ideological disappointment. He regrets 'the commercialisation of public services and the grotesque expansion of market forces into social relationships'. Furthermore, 'questions of status and self-esteem are now determined by the accumulation of material goods, not the maintenance of mateship … The politics of "me", the individual, replaces the politics of "we", the community'. The 'ladder of opportunity' took the aspirationals to Harvey Norman, not enlightenment. McMansion land is an artefact of 'the sorry state of advanced capitalism: the ruling culture encourages people to reach for four wheel drives, double storey homes, reality television and gossip magazines to find meaning and satisfaction in their lives'. In 2009, he scolded me for pointing to a 'false consciousness' amongst the working class, but he had himself earlier remarked:

> I wanted working people to enjoy a better standard of living, but had assumed [that] as they climbed the economic ladder, they would still care about the community in which they lived … especially the poor and disadvantaged. This was my misjudgement … they … left their old, working-class neighbourhoods behind and embraced the new values of consumerism.

Perhaps nothing inspires animus like an ideology rescued from the fires of disenchantment. George Orwell's *1984* concludes with a teary Winston finally realising that he loves Big Brother. This is, of course, at the end of a long and spurning road. It is just so for others with capitalism, that blank prairie of human possibility; at the end of the day, it is still a place to pull the creaking wagons up to, to water the weary horses, to let the children run free again … In recent years all signposts pointed to this golden grove of human achievement, but neoliberals never really told us what this marvellous 'go-no-further' place looked like. But now, Eureka. Mark Latham has found it. It's a shopping mall in western Sydney.

For me, at least, the escapism of the shopping mall, or the garden centre, or the retail whatever, is an utterly miserable view of human possibility. I don't mean that visiting and using these places (which I do) is wrong. But holding them up, and consumption generally, as the end point of human ambition is ridiculous. It's modernity in a cul-de-sac, oblivious to the dangers encircling it. We can do much better.

I will speak now about the possibilities which will emerge for our renewed attempt to realise the project of human emancipation we know as modernisation.

Ulrich Beck hopes for a 'reflexive modernity' that repairs some of the great mistakes of the past two centuries, the ones that largely explain our descent to the hellfires of economic and ecological crises. Industrial capitalism was premised on the 'excessive rationalisation' of modernity: that is, on far too great a faith in reason and rationality over other values, especially doubt. The final consequence of this was the creation of a risk society in the 20th century, of social institutions and practices blithely unaware of, and incapable of sensing the hidden contradictions and dangers in each new technological form. The development of nuclear power and weapons and the rising toxicity of industrial production and consumption were obvious ones. The steady accumulation of risks, both environmental and social, is one way of explaining our pathway to crisis and default. And globalisation put the process on steroids.

Beck reminds us that doubt is the forgotten twin of Enlightenment thought – we need reason *and* doubt. Attempts in the late 20th century to promote the 'precautionary principle' as a safeguard in institutional thinking were a belated realisation of the problem, but even they were brushed aside as prevarication by neoliberalism. In Beck's mind, 'reflexive modernity' would restore the twins, reason and doubt, to their shared pedestal, reinstating doubt and critical self-awareness as guiding values. In the new society to emerge after the fires of crisis this must be so.

From this point onwards, our first step in the pursuit of freedom must be towards a restrained and greatly transformed market. The neoclassical, neoliberal faith in the market as arbiter and allocator of all things is a perfect example of the excessive rationalisation that Beck speaks of. It is also a wretched assessment of human possibility, pushing critical decision making out of politics and into the sphere of mere exchange. The pure market is, after all, a machine that must decide things when democracy is pushed aside. The Australian city today, an artefact of neoliberal reform, is in many ways a stage for the play of brute power and money. The guardian state will have to put an end to this. What are we to do with markets afterwards?

First, there will be no chance of reinstating globalisation as we have known it. It ramped up overproduction until we finally overshot all possible limits. What we must commit to in a future world community is the principle of fair trade in search of co-operative mutual development. Hopefully the global co-operation, including in the areas of lawmaking and allocation, that was necessary during the crisis decades will have firmed the ground for a new model of international economy. It will be a world based on local determination of human and natural resources, far removed from the global web, with its fat multinational spiders, that binds us today.

This system would be founded on a new international commitment to the nurturing and renewal of human and natural resources. This means stronger ecological and social governance at the global level, from where national and regional development targets and plans can be identified. This would prevent nations from exploiting vulnerable resources (such as tropical forests) or destructive resources (such as coal reserves) simply because they can.

Markets will have been greatly restrained during the period of transition. Afterwards they must be overseen by institutional mechanisms that continuously reassess and ordain ecological limits. We should establish and maintain an Australia Plan, as I outlined earlier, as a basic national steering device that sets the course for safe renewable development. The plan must commit us to an economy and a society of reuse, repair and renewal. Its prescriptions will guide resource use and waste management. Its mechanisms could include a huge extension on all product warranty periods, to say 10 years, to screw down the instinct for overproduction. Natural limits and possibilities will of course shift as we adapt to the new climate regime and find more productive ways of living within its boundaries. To have a governance system much more in touch with ecology will require a huge investment in the research capacities of our universities and bureaucracies – of which more later.

It's not my intention in this book to outline an alternative political economy beyond this. It's enough to say that markets as we have known them for hundreds of years will have to give way to much 'saner', to employ Fromm's term, mechanisms for innovation, production and distribution. The expansionist reflex will have to be transformed into a quest for the flourishing of humanity and nature – which could take, surprising, not merely material forms. Think of an economy that directed productivity growth to the pursuit of time for creativity, including involvement in the arts and cultural expression generally by a much larger proportion of the population than is presently the case.

I think local ownership and determination will be the key to maintaining resources in natural balance, and with enough to spare to drive innovation and new uses that aim to improve our wellbeing. Our distribution and exchange systems may still look like markets, but they won't have the current fatal impulse towards blind expansion, and key resources will be owned and managed by communities not corporates. We'll have dispensed with the big institutional approach to infrastructure management, including its privatised versions, in favour of local, and usually municipal, management of natural assets and key services.

We should have left the queue and other heavy-handed allocation mechanisms behind, after the chastening time that had to follow our long binge on nature's credit card. It will be time to dismantle much of the centralised architecture of the economy that was necessary during the guardian state period in the interests of local autonomy and sustainability. But this will mean civic decentralisation, not privatisation. And all will sit within a regime of national and regional steering, including the Australia Plan, which must carefully work through the layers of government to provide broad but firm guidance to local effort.

I'll say a little more about how this comes together in 'The good city' (Chapter 14): how we must inherit and rebuild the lifeboats that brought us through the crises. My main ambition in this closing part of the book is to underline the necessity of a human needs model of economic development and to suggest that one great step towards its realisation must be a new focus on care as a driving social ambition.

Our new idea of freedom will be something that builds on and does not reject the gains we have made as a species since primitive times. I don't hold to the view that things are no better or worse than in earthier times, simply different. We have much to thank capitalist modernisation for. It opened the path to a new relationship with nature which has released us from the thrall of material necessity and made possible a new flourishing of our species. But now these gains must be built upon.

The two contradictions that will be rooted out of us are narcissism and materialism – the legacies of market-led modernisation. They are the juvenilia that have brought us to the court of environmental delinquency. As Fromm points out, a 'burning ambition for fame'[5] was not present in pre-capitalism and is not a hardwired feature of our species. Neither was an 'all-absorbing wish for material wealth'[6] evident in previous social forms. We have granted ourselves great and potentially useful material sufficiency and abundance, but have misdirected these gifts, creating the kinds of consumerist enslavement that Fromm and others observed with dismay in the postwar era.

When this society has passed into history there will be times of anxious transition ... followed, I hope, by open-hearted renewal. Then comes the possibility of development for us juvenile moderns. Conrad's 'fierceness of thwarted desire' might be harnessed to a new modernisation which aims to realise our species' potential. But first, it's back to some material basics. We must build a society that is unbending in its commitment to meeting the biologically given needs of *everyone*. This is to address the material deficits we know now as poverty, homelessness, ill health, and class injury generally.

From this base we must strive to fulfil psycho-social needs. This is to counter the malfunctioning we know now as sadness, madness and badness, the maladies that have thrived in the current version of modernity. Finally, we need to add to this project of liberation something not fully sighted by Fromm and his contemporaries – restitution to nature and its ecologies, the parent-force on whom we always depend. This means moving from a destructive adolescence to respectful maturity. It does *not* mean returning to the breast or the womb, as green idealism would have it.

The prized ideal of modernisation is 'freedom and integrity of [the] individual self'.[7] Having fallen out of nature we prodigals must find our way back on adult terms. For Fromm the way to species majority is relatively straightforward: we should unite ourselves 'with the world in the spontaneity of love and work'.[8] Freedom offers not indolence but a new unity of desire and effort. To gain it we must reassert the life forces that were anaesthetised by capitalism. Love is something we must pluck from the grasp of Hallmark and put (in)to work. How do we release love from the shackles of romanticism for the vital task of species growth? Can we have the freedom to love?

13

FREEDOM TO LOVE

Yes, there must be many bitter hours! But at last the anguish
of hearts shall be extinguished in love.

Joseph Conrad, *Under Western Eyes* (1911)

THE NEW SOCIETY TO EMERGE from the crises will not be perfect
– our humanity will guarantee that. Let me be clear that my view
of human improvement is pragmatic, not stringently moral. Values
certainly count for much. We need ethics and beliefs to provide value
frames and signposts that remind us of our better instincts, especially
our unique species capacity for fellowship, self-awareness and that
battered word, 'love'.

My emphasis here is not on any particular moral code, or on
urging us towards a 'good' society – in the happy-clapping sense of the
word. Of course I want us to have a fine society and in the next
chapter I'll consider prospects for a new home for *homo urbanis*, the
'good city'. But by 'good' I mean a good place to be, a place which is
good to us, not some sort of well-behaved, virtuous society – that will

ever be a fiction for us fallen humans.

There are two main ways I believe that we can learn and build from what we are about to pass through. First, as I've contended, we will have the opportunity to complete or at least advance our great species project of modernisation, seeking a deeper level of human fulfilment. Second, we must grasp the opportunity to remake our homelands and our life-worlds so that we are good to them, and they to us.

The two projects are of course related – by learning to see the Earth we will set the stage for a new realisation of the modern promise of freedom and fulfilment. There is no freedom or security for a species that continually sets fire to its own house, as we have done. The imperatives of repair and survival alone require that we make a good accommodation with nature. In an urban age, this means making a good city.

Humans pursued this project for millennia before modernisation; the two aren't necessarily related. So how will we chastened moderns achieve what our premodern 'inferiors' did with relative ease: living in a good relationship with nature? I'm not romanticising early history or its lingering forms. Premoderns changed the landscape and ecology, sometimes disastrously, but generally they didn't do this at dangerous scales. When things went wrong they never resulted in the planet-scale consequences that our mischief has wrought.

In this chapter I want to explore the first project. I believe the new society presents us with the opportunity to shift our social focus from complacency to care. There are many things we might do to reanimate the modern promise, but I believe this to be the most compelling and rewarding course. The practice of care will therefore be my principal theme. This movement of human sensibility and ambition then opens the road to a better relationship with our home-

167

worlds. The next chapter signposts this better place as the good city.

Let's go back to Fromm for a moment. Like Freud, he believed that human society has a powerful suppressing function. It puts lids on parts of our animal nature – our libidinal, feral drives – so that we can live communally. We get civilisation, but the price is restraint. On the flipside, civilisation frees and gives play to our creative potential as a species. And collectively we are much more powerful than we are independently. Only society can create the necessary creativity and energy to release new endowments and possibilities from nature. As Fromm writes: 'man himself is the most important creation and achievement of the continuous human effort, the record of which we call history'.

The nincompoops (US neoconservatives) who proclaimed the end of history in the 1990s had it half right. The crumbling of the Berlin Wall marked the end of the great 20th-century battle between socialism and the market. With the triumph of capitalism we really had reached a kind of dead end in terms of freedom; we were enslaved to consumerism and to an increasingly tyrannical and eco-cidal market. Where was that reservoir of creativity that our human society makes possible? More importantly, to what extent was it furthering real liberty, advancing the project of human self-realisation?

None of the ingenious things we have lately invented seem to have done this. The internet, computerisation in general, and the rise of global cultural industries have not led to greater levels of human fulfilment or liberation from work. Cultural critic Terry Eagleton provides a telling summary of the dead end we have backed ourselves into:

> One of the most powerful indictments of capitalism is that it
> compels us to invest most of our creative energies in matters
> which are purely utilitarian. The means of life become the end.

> Life consists in laying the material infrastructure for living.
> It is astonishing that in the twenty-first century, this material
> organization of life should bulk as large as it did in the Stone
> Age.[1]

Instead of deploying the vast resources created during modernity
to release us from drudgery, the market compels us to reinvest this,
endlessly, in the pursuit of more and more badly distributed wealth.
The great promise of the post-crisis age is the ending of this pointless,
ultimately dispiriting cycle. It is surely revealing that the maladies
that are on the increase in our late modern epoch are mental and
moral, not physical, indicating a failing of our minds and souls to
cope with this treadmill.

The crises to come will see history noisily restarting. Though they
threaten us, they also signal the reawakening of our creative energies
from a new social base. This base must be something other than
liberal democracy. Political theorist John Keane recently had this to
say about liberal democracy:

> This fine-sounding phrase is in fact living-dead. It is both a
> misleading descriptor of present-day realities and a worn-out
> nineteenth-century Orientalist ideal, at once in love with
> private property and contemptuous of ignorant, 'unwashed'
> people.[2]

The unwashed are the lead in saddle bags, the grumpies and the
defectives who slow the mindless quest for new wealth from old. For
a time in Australia, both right and left lampooned the One Nation
movement as a bunch of cantankerous hicks. But in reality, for a
moment the hidden injuries of reform had surfaced and spoken with
genuine anguish. Their cause was mocked as daft politics. I make no
case for One Nation: its ideas were largely intolerable to democracy.
Some denied even humanity. But its hurts were real and its sense of

justice keen. Liberal democracy created the conditions for this social wounding and then ignored – worse, scorned – those who endured it. The agonies of One Nation resounded with the cry of theft – of property, of life chances, of ways of life, of dignity – during the long march of neoliberal 'reform'. Amongst its ranks marched abandoned blue-collar legions, cashiered during the heroic years of structural reform.

The deeper message of this 'revolt of the margins' was that democracy seemed to have no claim, no cause for justice, over that arbiter of human life, the economy. It exposed liberalism as beginning at the wrong end of the scale, seeing the individual as the locus of human possibility and thus consistently underestimating and warping our potential to produce better ways of living, and thus real freedom within society.

Two forms of social compulsion must give way in our new attempt to achieve human self-realisation: worker ants sentenced to interminable 'organisation of things' and gamblers endlessly, recklessly investing gains, hard won and not, in the hope of greater and greater fortune. But what should we do with the petty compensations of consumerism and impotent individuality? How do we win back a mature relationship with nature, the parent from whom we were forcibly weaned? Fromm writes:

> There is only one possible, productive solution for the
> relationship of individualized man with the world: his active
> solidarity with all men and his spontaneous activity, love and
> work, which unite him again with the world, not by primary
> [natural] ties, but as a free and independent individual.[3]

In my mind, Fromm's powerful conjunction of three human ambitions – 'spontaneous activity, love and work' – are realised in one great endeavour, care. Care insists that we look beyond the miserable horizons of human interaction that we have set ourselves. As Fromm points out, the principal structures of our social relationships – consumerism, wage labour and liberal citizenship – share one great feature: 'mutual human indifference'.[4] During the great transition that the coming crises will force upon us, this indifference will be worn down by the need for solidarity if we are to survive. From this base we can envisage building a new society that replaces complacency with care.

Narcissism may be the guiding spirit of our self-absorbed age, but it is a cruelly twisted expression of the self-love we know is healthy and necessary. A commitment to care as a guiding, not to say overriding, social ambition would be the means for redirecting individual self-love to the species level. Despite our bravado, we are an intensely vulnerable species. Unlike other species, our young cannot survive unaided and we have immense and lately manifest capacity for self-harm and self-destruction. We also cannot subsist as 'breeding pairs' or warring communities: solidarity is a basic, if heavily repressed, necessity of our species.

Civilisations fight and collapse, to be sure, but the alternative is worse – our animal natures unrestrained, everyone belting each other's brains out. Curiously (or not), neoliberalism dragged us backwards towards the animalistic state and thereby reversed, rather than simply stalled, the process of modernisation. Lately apologists for this failed cause have tried to locate the capitalistic instinct in the DNA of our species. Capitalism may indeed prove to be the monkey business of our species, but this only confirms that it must be transcended as part of our evolutionary development.

Imagine then a new society where human care is a central ambition, not an afterthought, as presently (I'll come to environmental care in the next chapter). What would it look and act like? The premise is an

Australian society that has survived the trials and rigours of rapid and disconcerting climate change and a collapse in many vital bases and systems. It has discovered new ways of living well with much less, and thus of adapting to a much more challenging natural context. It has also learned, sometimes bitterly, to value solidarity and fairness and to cherish many intangible things, such as the quality of relationships and experiences. In some respects this society is more sinewy than its predecessors: conspicuous and gratuitous consumption are the stuff of history. Yet in others it displays softer embroidery, with a new recognition of human and natural vulnerability. By learning to see the Earth, and ourselves restored to it, we will have gained a new appreciation of nature's gentler textures.

Emerging from a war-like economy, we will possess the values and the means to build an entirely new materiality. Until the advent of neoliberalism, our political culture regarded the economy, and economic forces, as instruments for good, not ends or goods in themselves. To restart the cause of human fulfilment, the economy must be put back into its more traditional location: in the toolbox of human endeavour. The emergency period will have largely forced this to happen as natural threat and resource finitude overshadow political culture and become its leading concerns. The economy will be made to stand in the dock for a long period, accounting for its mischief under neoliberalism.

This is the very point of departure needed for a new social setting, that of care. We cannot give up labour, but we can imagine working with less alienation and for more fulfilling purposes than the acquisition of material goods and the enrichment of a few. We could transform and vastly improve our economy by redirecting a considerable proportion of its energy towards the practice of care. By ignoring care, we have generated huge material costs, dissatisfactions, losses and hurts, and they have usually been borne most by those with limited or no economic power. This cannot be discounted; it is why the army of One Nation was raised. It was a peasants' rebellion that went the way of most such eruptions from the powerless.

It is the poor and the excluded who bear the principal burdens of an uncaring economy; they tend to be revealed as physical and mental health morbidities, delinquencies, and bad, sometimes toxic, living environments. Yet every other social stratum, to the very top, is also carrying a share of the consequences of human frailty. These hidden griefs and grievances are papered over, with money perhaps, but nonetheless contribute to the great fund of human melancholy that our economy fails to value – or respond to – in any meaningful sense. The continuation of these malignancies is further testimony to the failures of our present view of modernity. We are hardly free when we cannot recognise, let alone repair, the ordinary injuries of human existence. We are further and perversely restrained when we add to the claims of nature the self-inflicted dangers of toxic industrialism. Both hurts are part of what Terry Eagleton describes as 'a general Western disavowal of uncomfortable truths' which betrays an underlying 'urge to sweep suffering under the carpet'.[5]

In Chapter 4 I summarised this situation as a crisis of underconsumption that shadowed our gross consumerism and overproduction. Our political culture, under neoliberalism, placed a permanent restraining order on any attempt to examine, let alone rectify, this perversity. A great arrival gift for us in the next world will be an understanding of our need to care for ourselves and our life-worlds.

In the next world we must create, with an eye as much to resilience as to progress, a society centred on care: our political culture, institutions and laws must mark it out as a central human expectation and thus an unbending obligation. Along with other guiding social values, care must direct the energies of an economy that is harnessed to the cause of human fulfilment. This means that a new social compact must be struck and enshrined. It must both guarantee a universal right to care and commit society to a great endeavour to meet that obligation. I've no particular thoughts on how that two-way compact will be established. Constitutional recognition could be the starting point. But there will be no substitute for a social and political culture that has reoriented itself towards care, seeing it not merely as an obligation to our human selves, but as an enormously enriching experience and practice. This means creating a much larger and prioritised space than we have today for the recognition and elaboration of care as a social ambition in our educational system and in our cultural life.

My vision is of care as a *nonspecific* concern of human society. By that I mean that I see it as a value that one breathes simply by existing in society, and constantly absorbs and matures with. It will no longer be a specialised concern of the 'caring professions' (who deserve our highest praise), or a 'community service obligation' to employ the murderously idiotic language of contemporary managerialism. Thirty years ago, Australian singer John Paul Young said that 'love is in the air'. Little could he sense the swirling miasma of neoliberalism that would soon suffocate his dream. In the next world his dream should be reborn in a society committed to Fromm's 'spontaneous activity, love and work' and where care is 'in the air'.

My proposal is that at least a third, or as much as a half, of our paid work effort should be redirected to the provision of care. Human frailty takes ever-evolving forms. In the first brutal phases of capitalism the infirmities were physical: injuries and diseases caused by industrialism and colonialism. By the 20th century our maladies, especially cancers and cardiovascular diseases, betrayed our increasingly toxic and ill-nourished society. Now, I suspect these great diseases are giving way to remarkable medical advances, at least in richer nations, leaving a world of mental sickness and heart sadness. This is an unfortunate situation as we enter a time of stress and threat. We cannot yet guess the frailties of the next world. We may well emerge from the time of trial with new wounds and grievances.

I've said that care must be a guiding value for the economy, but it must also occupy a large part of our systems of 'material organisation'. This will have the colossal added benefit of decarbonising our economy. Care and education use less energy than most other economic activities. It's a revealing corollary that the preoccupations that consume most energy are probably the least productive in terms of human nourishment and natural renewal.

Importantly, we will not allow monetary profit from care. The newly localised economies I foresaw in the previous chapter will have as part of their charge the planning and delivery of care. I'm not advocating American-style municipal fiefdoms, where local governments get on as best they can without much direction or support from above – meaning the rich communities do well and the rest languish. The planning and funding of care will begin with the Commonwealth, and proceed through the states or whatever regionalised governance might one day replace them. It may be that the climate crisis will force the guardian state to restructure the boundaries of federation around natural defence lines, such as catchments or bioregions.

So a major part of our economic effort will be directed to care, and the rest of the economy will be made much less indifferent to the values of humanity and nature. This means a vaster share of community wealth than is presently the case must be dedicated to nurturing, instead of to preening ourselves. The shift to community steering and ownership and the imposition of hard and binding ecological limits on markets (or the trading spheres that will replace them) will do much to reduce the malign complacency of the mainstream economy. If we end the suicidal compulsion to growth and learn to value human experiences and fulfilments – relationships, culture, ecology – we'll have fewer qualms about the 'cost' of care.

One necessary role for the Commonwealth will be to ensure that the atmosphere of care is maintained and that resources are deployed fairly amongst communities and between regions. I'm attracted by the example of some European nations, which have social service as an option for young people who would otherwise be drafted to the military. I've no real idea how successfully this works presently in such countries. I believe, however, that in a society whose culture is care-focused the drafting of the young to such work would be a source of enrichment for giver and recipient. What better preventive medicine for the agonies and angsts of youth than universal human service? If we are still ageing as well as we are now, at least in Australia, there will be another population group available for voluntary service. The grey reserve army should be recompensed for service to supplement whatever form of pension we maintain.

Care in the coming world must extend to many different – though overlapping – realms of human need. I won't try to nominate or address them all here. As I pointed out in Chapter 4, all of us at some point in our lives will be part of the great and still largely unheralded care enterprise. Whether through age, disability, sickness, sadness or fatigue, each of us will encounter the radical limitations of what it means to be an individual. However, many of us will not be called upon to care. For a start, our heartless society tends only to care in the last instance, when the injuries of human existence become too acute or too burdensome to ignore. Worlds of need are ignored: the isolated, overworked contemporary family comes to mind. Then our residualised 'duty to care' is largely outsourced to others – the permanently stressed state and voluntary sectors, and the emerging feral theatre of commercial care. If you are still looking for evidence of freedom half-born, go no further. Consider how much of our creativity and spontaneity, our potential for happiness, and our physical energy is sapped or spoiled by the absence of care.

The great enterprise of care I foresee in the next world will begin with the family as a central concern. It must, because we have left this most vital social organ exposed to the vicissitudes of an aggressive and fluid market for too long and at great cost. Some of this fluidity has been hugely beneficial, allowing the opening out of the family to accommodate many different forms of living. Its recent nuclear form was a historical aberration and we should welcome most of the new ways in which family life is conceived and practised. Some family forms are problematic. I hold grave doubts about the family that outsources the raising of children to commercial 'care', but I recognise that for many this is a necessity, not a choice.

American psychologist Mary Pipher considers the injured modern family in her 1996 book, *The Shelter of Each Other.*[6] Pipher explains the family as an isolated, colonised, largely defenceless unit – beset by the hostile forces of consumerism, adult overwork, injurious technology and culture. Families struggle and increasingly fail to screen out the harm from the good in these influences, which have found new ways to penetrate the ordinary human life-world. Rapidly evolving media, internet and entertainment technologies have tunnelled into and undermined the basis for a secure family life. A whirling mix of stress, overconsumption, violence and addiction now envelops our families.

The debilitated state of our families highlights three needs which must be addressed in the next world. Mary Pipher sets tasks for each. The first is to reset family life within a resecured and securing communal landscape and to thus transcend a therapeutic culture that sees human injuries of the mind and heart as individual or even family pathologies. She is critical of the pathologising work of her own profession, which has tended to mask the social and wider origins of individual and family breakdown. The second task is to provide what she terms meaningful, ethical work, freed of the alienation and mindless compulsion of today.

The last need requires a recasting of human communication by protecting, in some instances unplugging, our homes from unmediated commercial technologies (the internet, gaming, telephony, electronic media). Part of Mary Pipher's response is to still the world, to protect families (let's say households) by recreating what she terms 'houses with walls'. She describes a set of protective strategies, such as creating hallowed moments and places in which time is stilled and the world pushed aside. The next step is to connect new 'sheltered places' to form a larger protective sphere, of 'sensuous' (I can see it, smell it, feel it), caring communities.

Our relatedness to others and the wider world must be mediated by values and priorities that are physically close to us, and over which we have some hope of influence. We can find this in the simple common purposes that healthy neighbourhoods practise: protecting amenity, especially in communal areas, maintaining neighbourliness, guarding against hurtful intrusion, and welcoming strangers. This returns us to the first need Mary Pipher identifies – for repaired community. A culture of narcissistic focus on self must be replaced by communities of self-reflection and support, set within a climate of social solidarity and resilience.

Our commitment to a great care endeavour will shelter the family and restore its sheltering capacities in a world that will always, despite our best efforts, put weights on human existence. Our first and most egregious failing today is to leave the important task of rearing children to families that are already stripped down, battered. In the next world, an adult carer in a household with children will be socially supported. The remaining need for care will be met in community enterprises that provide a point of contact and provision for all family services.

I envisage a more fulfilled but not perfected species. Frailty, mortality and the inescapable 'burden of existence' will, I believe, always mark our species. The new world of care will do much to meet the three needs outlined above. Do not discount how much of our hurts and harms arise from, or are worsened by, loneliness and physical isolation, especially in large, even 'compact', cities. Esteemed disability advocate Rhonda Galbally recently reported on a study which found that 'people with a psychiatric or intellectual disability will have a negative social experience within fifteen minutes of leaving their home'.[7]

With a vast new amount of labour power available, we should be able to create supporting teams and communities for every care recipient. The needy will no longer mistake a voucher system of minimal support for 'independence'. Autonomy and self-realisation to the largest extent possible will be gained through interdependence, not independence.

Institutions need not return. The consumerist ideal of 'care in the community' will be overtaken by caring communities. We won't parachute the needy into indifferent neighbourhoods, but will instead carefully place them within networks of carers, teachers and friends. The sense of exceptionalism that colours the contemporary practice and experience of care will recede in a culture which takes the limits of individualism as given. In every community economy, human needs and needy humans will take precedence over other material possibilities. Good riddance to all that!

Learning to see the Earth means a constant state of reflection upon our ever-evolving world. The lesson is never over. A culture attuned to care is much more likely than others to maintain this commitment. Care is the antithesis of complacency and, ultimately, of narcissism. It demands that we see ourselves as needy, at best temporarily capable, and bound indissolubly in networks of interdependency with humans and nature. If the carer occupied a valued role, not a marginalised one (as presently), we would be less prone to the vanities that have caused so much misery in modernity.

We have nothing to lose and much to gain by embracing the national project of care in the next world. We will not squander our precious individualism then, because we have precious little of it. The ruse of liberalism is unravelling. Self-hood has been steadily thieved from us through work alienation and consumerism. The 20th century, the 'corporate epoch', offered only deeper levels of insignificance to the individual. Before its halfway mark, Erich Fromm was moved to conclude that 'The free, isolated individual is crushed by the experience of his individual insignificance.'[8]

By devoting ourselves to 'spontaneous activity, love and work' we can hope for higher levels of fulfilment and self-realisation. For the needy, the prospect of a caring society promises only better things. As Rhonda Galbally observes, 'Australia is not a liveable country for people with a disability.'[9] This surely speaks for every identity which departs from the consumer-citizen ideal. At some point in our lives it speaks for all of us.

The market, as we know it, is a recent human creation. There are those who would naturalise this artefact; they claim to find it in the behaviour of forest-dwelling primates. And yet the law of the jungle has proven the enemy of the forests themselves. We are the only species that has set itself so firmly against the renewal of nature. The 'rule of the market' is human monkey business.

The bitter hours of default approach. The anguish of human hearts began long ago. Modernity, now doubly censured, faces its greatest test. And yet the judgement need not be what we fear most – a great mortification of the species. The clemency of care beckons. In a trial that results from own actions, we are granted one favour: the chance to write our own verdict. May it read, after Conrad, 'At last the anguish of hearts was extinguished in love.'

14

THE GOOD CITY

A place where I forgot to be happy
Because I was

David Malouf, 'Out of Sight' (2007)

THE CITY MADE US FREE; it was our first modern home. Julianne Schultz reminds us of the old German injunction *Stadtluft macht frei* ('City air makes you free').[1] This same people turned the saying into an abomination – *Arbeit macht frei* ('Work makes you free') – for their industrial death camps. Work, stripped of spontaneity and love, and shackled to murderous purpose, was the motive of those deathly cities. The homicidal factories of Nazism scorned the great modern hope that humans in close society could create freedom.

We have frequently used the city to erect structures dedicated to conceit and avarice, not freedom. Our towering Babels have been monuments by which we have thumbed our noses at nature. James Lovelock makes the point with characteristic gentleness, observing that 'human animals building and living in our city nests [are] slowly severing contact with Gaia' and that they (we) are 'ultimately in danger of becoming the real and predatory aliens on what had been the planet of our birth'.[2]

A crumbling, over-extended empire cannot offer freedom to anyone. The two dimensions to the modern quest for freedom, reason and doubt, must continually question each other. Industrial modernity dethroned doubt and built towers to reason. Our eco-cidal cities demonstrate the consequences of excessive rationalism. We will learn to doubt again the hard way, but only by suffering the inclemency of nature wronged. Humanity must endure a sentence of stress and privation, and our cities will be the lifeboats in which we will be interned for a time. The city will carry us through these storms ahead but will, like us, emerge battered and barnacled. All monuments to the 'suicide instinct' of modernity will then have passed into history.

Forty years ago, French geographer Jean Gottman observed that massive industrial urbanisation 'could hardly have happened without such an extraordinary Promethean drive'.[3] The new worlds of North America and Australasia, where nature lay unclaimed and uncosted, seemed perfect places for the 'Promethean endeavours that had long been confined to the dreams of European people'.[4] This 'immense experiment' was later extended through globalisation to the remaining unexploited parts of the globe, plastering it with that new testament to industrial ambition, the 'megalopolis'.

The mixed human fortunes of the experiment cannot be denied: untold lives were tossed into the furnace of brutal urbanisation, but in time its fruits improved the material conditions of many. Urban civilisation created opportunities for new forms of human learning, expression and ambition. The suburb placed a restraining order on the most vicious forms of urbanisation and provided the setting for a great improvement in wellbeing that cannot, alack, be extended universally without great cost. Now, in an era of failing industrialism and toxic growth, the new and developing worlds continue to create these crowded cities. Cyclonic urbanisation has battered the city ideal. The new dreamers write pulp fictional accounts of human escape to space colonies.

Meanwhile, back on Earth, the limits and paradoxes of an urban model that promised everything and delivered much are steadily being revealed. There is no 'boundless vista of unlimited resources for an affluent society'[5]. The horizon of failure and threat looms. Some sensed this coming even at the height of the carnival. Jean Gottman wrote in 1961: 'As the frontier becomes more urban in its very nature … the vultures that threatened Prometheus may be more difficult to keep away.'[6] In the next millennium, *homo urbanis* may ponder these remarkably prescient words. Our frontier has moved from raw nature (the wilderness) to the city. We need cities that provide good prospects for human fulfilment, once we are ready to take the road of modernisation again. What are these good cities?

Since Augustine's *De Civitate Dei* (City of God)[7] there have been many attempts to describe 'the good city'. Today, revealingly, the 'City of God' is a slum in Rio de Janeiro, a city of wild social extremes that typifies the post-neoliberal future we may be drifting towards. Augustine wrote his tract at a time of great turmoil, resounding with iconoclasm, in the early 5th century. Rome the inviolable had fallen to the Visigoths in 410. Amidst the broken idols of earthly rule, Augustine urged all eyes to heaven and to a higher, more enduring sense of human purpose.

We can take much from this today, in metaphor and as inspiration. The Visigoths were surely rough nature reclaiming human society, pulling it back to Earth from the peaks of 'over-civilisation', hubris and vanity. And yet there could be no return to grubbing. From the ashes of folly emerged a new design for human endeavour, the City of God. Human history, for Augustine, is a conflict between desires, for the City of God and the City of Man. We know this struggle well in modernity, where the suppression of animal desires has been seen as necessary for the releasing of the possibilities of human society. The City of Man is a confederacy of brutes: a society given over to species chauvinism and to the satisfaction of natural urges – in two words, Prometheanism and materialism. The City of God, re-read today, is a society that holds to hope, to the idea that human fulfilment means more than the half-born freedoms of consumerism and liberal democracy.

If hope sounds musty and anachronistic today, it is because we have lost our understanding of why we set out on the modern road in the first place. The only possible endpoint of a 'postmodern' age is a place without hope, a place that Augustine and the Church would call despair. I think we underestimate how close we are to this place. The coming crisis might be the one thing that steers us somewhere else. The good city is our chance. We can leave this mullock heap of industrialism. We have brains and a unique species capacity for critical reflection, and we should take this chance for freedom; it may be our last.

The question of the good city is a compelling one now that we are an urban species. We must reach its shores after the storms of change. How do we best conceive the good city? We moderns have been transfixed by the notion of a perfectible city. The modern urban pattern book started with the star-shaped fortress towns of the Renaissance, and proceeded through various technically ideal forms – the model factory town, the garden city, the city beautiful, the new town ... Under neoliberalism, all these fancies were boiled down to the master-planned product that studs the outer regions of our cities today. As these indicate, perfection was not achieved. The successful models are few, though we can be very proud of Canberra, a fruit of modern ambition that is maturing very well. Visions like the garden city ideal that thrived in the early 20th century may be impractical, but they can exercise a good influence on urbanism. Australian suburbs, our great heartlands, were improved by these guiding models.

We erected the standard of freedom in the cities – they meant liberation from grubbing in the earth and from the tribalism, in many forms, that had bound us. The cities of the Renaissance and the Baroque celebrated and experimented with these new ambitions. With capitalist industrialism, the flight to freedom sought to scale much greater heights, beyond the reach and presence of nature itself. Now we are falling backwards, like Ancient Rome, from the upper reaches of this desire.

Nature is, if nothing else, the rule of the temporal, and the city is humanity's attempt to depose time, its oldest foe, by rationalising nature. Everyday urban life can be understood as a great act of mutiny against the natural forces that govern all materiality. And yet this urban rebellion produces not the reign of pure reason but an unruly disruption of nature, and the speeding up of time. Consider the mutinous storm of industrial modernity and its rapid perturbation of the Earth's climate. Blueprint plans, ideal dioramas, model urban villages, all the idols of rebellion, are now destined for oblivion.

The most recent urban idol, the compact city, comes not from the green earth, as it pretends, but from the pattern book of industrialism. It is a vision of closely crowded urban humanity, a tall city casting beams of reason on distant nature. Science doubts the ecological beneficence of the model, and we all may question its towering ambition. It is a new instance of 'excessive rationalisation' denyng the impulse of doubt.

The real project of urban modernity is best defined as a set of relationships and values that encourage human and natural flourishing in urban society. These urban values are the ideals which must guide us through the trial ahead: justice (meaning equity, not just the rule of law); modesty (meaning restraint as the necessary safeguard of civilisation); and solidarity (meaning the recognition and nurturing of human interdependence). These are the values of the good city. They must serve a higher necessity: the cause of nature's renewal. The good city will be an artefact of human will that constantly replenishes its natural foundations.

The insights of modern, especially evolutionary, science are guiding posts for our urban future, because the city is properly conceived as a restless social-ecological system. It is not a set of fixed dioramas (suburb, apartment-scape, knowledge city, urban playground) but a flow of nested processes that continuously evolve. The imprint of evolution is present at every level, beginning with the growth and development of human subjects. Evolution and adaptation are endemic, not optional. The climate crisis already breaking over us merely underlines, dramatically, the necessity of constant urban adaptation. The other bases of human existence will also evolve, including some that carry few obvious traces of our handiwork.

Almost every natural process and resource is now caught up in some way with the development of humanity. Mostly our influence has had harmful consequences for the Earth's systems. It need not be so. The proliferating and excruciatingly noisy crows in my neighbourhood (they caw! as I write) are joined indissolubly with human urban life. We have expanded their food supplies through urbanisation and have reset the course of their evolution, and thus also of those they feed on. In the good city we will be aware of such things. Good cities will join non-human species and nature generally in a project of mutual evolution that sets its sights on peaceful co-existence as much as on the liberation of new human possibilities.

Evolution and resilience are watchwords of the good city. The concept of 'end state' belongs to despair, to time finally stopped for eternity. There is no end for the good city which takes the project of continuous human realisation as its guiding purpose. Its premise is evolution, ceaseless in motion and restless in form. The 'heavenly city', as its name suggests, lies beyond the horizon of human consciousness, in death. The living can hope to inhabit the good city. We must accept the inevitability of evolution, and the necessity of adaptation. This means that there is no definitive blueprint of urban 'goodness'. We need to hope not for a final model of stabilised optimality (the urban village, the compact city, the eco-village) but for a continuously adaptive and resilient urban system. The goal of urban planning should be to maintain and safeguard the human place in evolution, because that place is urban.

Urban planning has long sought to defeat disorder – of the market, of venal human intention – but disorder should not be confused with evolution and its offspring, complexity and uncertainty. Planning in the next world must steer change, mould it, in search of urban resilience. As part of this it must preserve, in a context of shifting possibilities, pathways back to human needs, including shelter, solidarity, stimulation and self-determination.

The task of social steering will not be left to planning, as there are no simple 'spatial fixes' for evolutionary dilemmas. Planning cannot recognise, let alone check, the underlying causes of environmental and social threat in market societies: the tendency to uneven social development and to overproduction inevitably flow from the 'growth fetish' of contemporary political economy. Resilience and fulfilment cannot be achieved by shifting the pieces on the board; it requires constant adjustment of the game's rules, aims and tactics. The shift to community economies and the great new enterprise of care are the building blocks of resilience in the next world. Planning must work towards their realisation by localising urban life and nurturing the activities that support the renewal of humans and nature.

This is not the return of the stultifying village, nor of forced neighbourliness and daily work in the (suburban) fields. We have no map to show us how to proceed, but this much is obvious – human effort will have to be sourced and expended much more locally than now because all energy will need to be nurtured, conserved with care and ingenuity, and expended with fairness and accountability. A great and ultimately destructive energy imbalance will be recalled as a defining feature of industrialism, and of neoliberalism in particular. Material energy was destructively sourced and used, and human energy wasted on meaningless 'material organisation' (to recall Eagleton). The planners of the good city will be conservers of energy, human and natural.

Our physical designs in the good city will reflect as much as shape our political and social designs, many of which will invite critical reflection, not celebration. Too often grand design ambitions in our cities are indistinguishable from the play of money and power. Our CBDs are littered with wasteful follies that mark the courses of power, not of human creativity. Meanwhile, in the suburbs, as musician Dave Warner once said, generations of small human ingenuities have shown how design can wear a humble suit and yet serve the cause of human realisation in quiet ways. We can of course do much better in both landscapes.

Good city design will recognise the power of built environments to shape human societies and their ecologies. A high-density urban village might be conducive to social interaction and happiness – but not if it's badly designed, car dependent, congested, polluted and home to an elite 'demographic' who want peace and privacy, not encounter and surprise. And a low-density suburb might have a relatively modest ecological footprint, if householders restrain their consumption of goods and services, use cars only when necessary, grow food, collect and treat water, and shop and socialise locally.

But all this is a contemporary view. The design task is at best a moment in the larger clockwork of urban endeavour. Evolutionary forces, including the all-too-visible human hand, will set us on a course of continual redesign that will disrupt and corrode any attempt to establish some permanent, eternal plan. After the crisis, a changed climate will shape our range of urban configurations. Defensive or 'contained' cities may be needed to keep a disorderly climate at bay and to preserve vital resources, especially water. Or perhaps our crisis-honed insights and technologies will yield a more dispersed settlement pattern. We cannot know the urban pattern in the next world, but we can commit to a search for good, not miserable, forms.

Resilience is a vital footing of the post-crisis urbanism that will lead us towards the good city. It can be restated as one word – 'equity', meaning fairness. The link word is 'solidarity' – we must be solid and stand together to refuse individualism. There is ample historical and social science evidence to demonstrate that equity and social solidarity are positively correlated, and that both generate high levels of collective welfare. This received wisdom is strengthened by mounting evidence that equity restrains environmental degradation and reduces social exposure to ecological risks.[8]

The deterministic cloud that has settled over Australian urban debates seems to have stifled commitment to equity and solidarity. We had it earlier in the writings of Hugh Stretton, Patrick Troy and Leonie Sandercock. Some environmental interests see equity as a subsidiary or secondary priority in the face of ecological threat, but fairness is in fact a precondition for sustainability. The radical and unworkable imposition of towering densities urged by some, especially in the urban design fraternities, shows their impatience with the cause of equity. The proponents of unfettered urban and suburban growth sometimes pressgang the language of equity to their cause by invoking the collective dream of home ownership. In fact they mask the rule of money and power in urban land markets and deny the sustainability threat that looms over urban Australia.

Australian urban studies would do well to reawaken the interest it once had in urban equity. It will be a virtue forced on us during the time of trial and pre-emptive thinking would be helpful, to say the least. The evolutionary view demands a vision of urban equity fitted to a changing climate and a shifting resource base. In the 20th century, suburbanisation was the means by which a good life was made possible for most Australians. This, in conjunction with motorisation, must yield to a new model of urban justice. However we do it, there must be space for everyone to pursue the project of human realisation, which itself will take new as well as old forms, including the bearing and rearing of children, education, and a new interest in the cause of nature's renewal.

Apart from this, justice will be served and signalled by the massive accommodation that the good city will make for care. A new *stadt luft* will circulate within the good city, an atmosphere of care that fosters the capacities and hopes of every citizen. No longer deprived of oxygen, the enterprise of care will itself speak much more rationally than it does now to our desires for realisation and freedom. Care will be the path to freedom for all, not today's cul-de-sac of failed human possibility. In a re-localised economy and city, much of the energy previously expended on frictional movement, 'material organisation', will be redirected to the practice of care. Localised economies will employ locals, communities will find communion and inspiration in care. But by then this will be an ordinary, not remarkable thing.

Finally, the call of solidarity must awaken and raise up again a landscape submerged during neoliberalism: *terra publica*, the spaces, places and activities that belong to us all. The instinct for private property implanted during capitalist modernity will have been greatly blunted by the time we reach the next world. We will have passed through a long time when such self-interest will have been restrained in the interests of collective survival. We will be free again to feel emancipation in less proprietary ways of living. The community will retain, and when necessary exercise, the final ownership and use of property. Leasehold, as we have long had in Canberra, should be the exclusive basis for all land use.

There will be no abrupt movement from private to public space. The public realm will not struggle for survival against giant corporates in an economy reorganised at the community level. *Terra publica* will be a patchwork of public spaces, places and endeavours that affirm and nurture social solidarity and individual expression. Part of the enterprise of care will be dedicated to the cultivation of *terra publica*, including the new food-growing spaces and wild areas that will flourish in the good city. Care will extend to many other practical tasks of social and ecological renewal. As part of its leasehold responsibilities, the community will maintain the dwellings of those who need help.

The good city will not be a society of saints. Nor will it embrace secular ideals of perfection – the dour market town of chaste liberalism, the consuming circus of neoliberalism. It also will not be the rough-hewn utopia of 20th-century socialism, full of cabbage queues and watchfulness. The good city will transcend industrial modernity to realise a new modern possibility. It will replenish and extend our hard-won gains of civil and personal freedoms. The good city will preserve our power to transform nature to free us from material necessity, but this power will be put to a better end than money chasing money. The obsession with accumulating things will have passed. With time to care we will realise many new human possibilities of expression and creativity. Care will be an ordinary thing, but still hard work. Power will be an ordinary thing, dispersed to communities, but still our greatest venality.

The good city will be a better place than now. Not for its virtue; this will wax and wane. Goodness will emerge from a new reciprocity between humanity and nature. We will care for this city and it for us. This will mark a new maturity in human achievement. We will have survived the slaughter of value that we knew as industrialism. The great trial that followed will, eventually, have ended well. As we make our exhausted way to new shores, we will have finally learned to see the Earth again. We will have done everything the hard way, *via humanis* perhaps. The new Australia we re-inherit will be, in the words of musician David McComb, a 'beautiful waste'.

We ate the fruit of knowledge. We cannot undo its work. We cannot go back to the first light of purity. There is only the way ahead. Eden must be renewed through our own cultivation. *Homo urbanis* should find its good home in the city. If we really love freedom, the taste of the fruit, we must free the work of love. This will not happen until we rouse from the slumber of failed modernity. Scorned nature clamours in our sleep. The apple cast a spell: the hypnotic brake on our senses we called individualism. That dream is reaching its improvident end. We wake in a time of vast possibilities. In the next world we could be true monarchs of the beautiful waste. Or we could be dust.

AFTERWORD

I close this book, observing through the office window my son in our forested suburban garden. Julian's happily in his own world, slowly bouncing on a trampoline in a meditative state. My wife sits nearby reading at our garden setting, at peace in his calming play. For years it has been his practice to enjoy in this most modern way the peace and amenity with which we have been blessed. He's a budding teen now, so the enjoyment is enriched and complicated (as it must be) by the addition of an iPod and feverous music. On the tramp his gawky teenage limbs are forgiven and achieve a kind of balance not possible on dry land. His younger sister, Alison, is busy with some important little project in the house, and safely so. For her the garden is still a giant antechamber to an unknowably large world. She's learning to make her way to it, but she still needs us with her. Our little suburban world traces the wider architecture of human development.

Not so long ago other peoples ranged over the Seven Hills of Brisbane where we live. The Turrbal people were the first owners of the Brisbane Valley. Their children must once have wandered through the bush outside my window. We took and subdivided their lands to make our farms, our suburbs, and our entire modern landscape. We own the land in a capitalistic way; will we own the consequences of our misuse? Is this legacy our property too?

A civilisation preceded us. Before there were suburbs and cities and rural areas, there was the land and its peoples. This world and its inhabitants were savagely suppressed. Its adults were shot down and

194

its children were 'brained', clubbed on their heads until dead. For decades our forebears unleashed the Native Police to clear the way for orderly development. We have this barbarism at the core of our development history and we should not forget it, or the souls who were slaughtered in its quest. To our great and undeserved good fortune, the civilisation we set out to annihilate has survived. They came through our slaughter.

I return my gaze to my son. He bounces on, oblivious to me, in his hybrid world, nature with a soundtrack. Julian doesn't yet know the murder and more that made all this possible. We've had leaders who would prefer our young, indeed any of us, to never know or acknowledge this. He doesn't yet sense the hard road ahead, and I am thankful for this. It worries me to bits that he and his sister must make their way through a troubled world.

What pleases me, however, and gives me hope, is that he and his generation seem less captive of the ridiculous braggadocio that has coloured the neoliberal era. They might be the generation to make the change that is achingly needed, to a new Australia. They might be the first Australian moderns who dare to admit that we got things wrong, that we badly misread this continent and its ecologies. They might be the ones to ask its original owners how to live peaceably with an offended land. Hope takes surprising forms. That's what it looks like to me this summer evening as we drift towards the next world.

NOTES

Chapter 1

1 M. Pusey (2008) 'In the wake of economic reform ... new prospects for nation-building?', in J. Butcher (ed.) *Australia Under Construction: Nation-building – past, present and future*, ANU E-Press, Canberra, p. 27.

2 B. Gleeson (2006) *Australian Heartlands: Making space for hope in the suburbs*, Allen & Unwin, Sydney.

3 G. Davison (2004) 'Conversations about suburbs and communities: the new Griffith Review', *ONLINE Opinion*, posted 12 January 2004, accessed at: http://www.onlineopinion.com.au/view.asp?article=1204.

Chapter 2

1 J. Birmingham (2009) 'The coming storm', *The Monthly*, no. 46, June, pp. 22–30.

2 D. Marr (2007) 'His Master's Voice: The corruption of public debate under Howard', *Quarterly Essay*, no. 26.

3 C. Jackman (2003) 'Chloe: A victim of life on the fringe', *The Australian*, 15 November, p. 1.

4 A. Sandy (2009) 'Alpha accommodation: Where police fear to tread', *Courier Mail*, 29 May.

5 Australian Senate (2007) *Inquiry into Australia's future oil supply and alternative transport fuels: Final Report*, Australian Senate, Canberra.

Chapter 3

1 A. Ferguson (2009) 'One-third of nation at risk of loan default', *The Australian*, 7 September.

2 K. Knight (2009) '60,000 hungry', *South-East Advertiser*, 11 November, p. 1.

3 G. Megalogenis (2006) *The Longest Decade*, Scribe, Melbourne.

4 C. Martin (2009) 'Let's not forget those still living in poverty as the economy picks up', *The Age*, 15 October.

5 A. Sandy (2009) 'Fear for kids in park hell', *Courier Mail*, 29 June, p. 7.

6 Knight, '60,000 hungry'.

7 Department of Planning and Community Development (Victoria) (2009) 'Community and not for profit organisations' capacity to manage the impact of the Global Financial Crisis 2008/2009', accessed at: http://www.dvc.vic.gov.au.

Chapter 4

1 N. Cica (2007) 'On the ground', *Griffith Review*, no. 15, p. 178.

2 D. Harvey (2009) Address to World Social Forum in Belem, 29 January.

3 M. Clayfield (2009) 'Tax rise to bring a smile', *The Australian*, 28 July.

4 P. Pape (2009) 'Green shoots or toxic haze?', *Courier Mail*, CM2 Supplement, 18 May, p. 36.

5 F. Stanley, S. Richardson & M. Prior (2005) *Children of the Lucky Country?*, Macmillan, Sydney, p. 52.

Chapter 5

1 'Fabled sea route unfrozen', *Courier Mail*, 13 September 2009, p. 49.

2 P. Spearritt (2007) 'The water crisis in Southeast Queensland: How desalination turned the region into carbon emission heaven', in Troy, P. (ed.), *Troubled Waters: Confronting the water crisis in Australia's cities*, ANU E-press, Canberra.

3 M. Latham (2009) 'Actions contradict polls', in E. Beecher (ed.), *The Best Australian Political Writing*, Melbourne University Press, Melbourne, p. 211.

4 R. Reich (2008) *Supercapitalism: The transformation of business, democracy and everyday life*, Scribe, Melbourne.

5 D. Harvey (2008) 'The right to the city', *New Left Review*, 53, p. 24.

6 Ibid.

7 D. Meadows, D.L. Meadows, J. Randers & W. Behrens (1972) *The Limits to Growth*, Universe Books, New York.

8 T. Friedman (2009) 'The inflection is near', *New York Times*, 7 March.

9 R. Wright (2004) *A Short History of Progress*, Text, Melbourne, p. 129.

10 Megalogenis, *The Longest Decade*.

11 E. Fromm [1942] (2009) *The Fear of Freedom*, Routledge, London.

Chapter 6

1 N. Brown (2001) *Children of the Self Absorbed: A grown-up's guide to getting over narcissistic parents*, New Harbinger Publications, Oakland CA.
2 A. Manne (2006) 'What about me? The new narcissism', *The Monthly*, June, no. 13, pp. 30–40.
3 Ibid.
4 Ibid.
5 B.J. Gleeson & N. Sipe (eds) (2006) *Creating Child Friendly Cities: Reinstating kids in the city*, Routledge, London.
6 Cited in F. Metcalf (2009) 'Facing up to a new suburbia', *Courier Mail*, CM2 Supplement, 24 August, p. 30.
7 R. Evans (2007) *A History of Queensland*, Cambridge University Press, Melbourne.
8 J. Borger (2008) 'UN chief calls for review of biofuels policy', *The Guardian*, 5 April.
9 Manne, 'What about me?'.

Learning to see the Earth

1 R. Evans (2007) *A History of Queensland*, Cambridge University Press, Melbourne, p. 140.

Chapter 7

1 Cited in *The Australian*, Inquirer, 22–23 August 2009, p. 26.
2 J. Hansen (2006) 'Climate change: On the edge', *The Independent*, p. 1.
3 J. Lovelock (2009) *The Vanishing Face of Gaia: A final warning*, Penguin, London.
4 Ibid., pp. 19–20.
5 Ibid., p. 87.
6 Ibid., p. 4.
7 C. Forster (2006) 'The challenge of change: Australian cities and urban planning in the new millennium', *Geographical Research*, 44(2), pp. 173–82.
8 G. Monbiot (2008) *Bring On the Apocalypse: Six arguments for global justice*, Atlantic Books, London.
9 J. Houghton (2003) 'Global warming is now a weapon of mass destruction', *The Guardian*, 28 July.
10 Lovelock, *The Vanishing Face of Gaia*, p. 160.
11 H. Stretton (2005) *Australia Fair*, UNSW Press, Sydney.

12 Ibid.
13 S. Biddulph (2007) 'The party's over and Liberals will soon be history', *Sydney Morning Herald*, 29 November.
14 M. Peel (2007) 'The inside story of life on the outer', *The Age*, 16 September.

Chapter 8

1 Cited in D.J.C. MacKay (2009) *Sustainable Energy – without the hot air*, UIT Cambridge, Cambridge, p. 248.
2 D. Horne (1964) *The Lucky Country*, Penguin, Melbourne.
3 Forster, 'The Challenge of Change', pp. 173–82.
4 Horne, *The Lucky Country*, p. 16.
5 J. Hirschhorn (2005) *Sprawl Kills*, Sterling & Ross, New York.
6 E. Farrelly (2007) 'Our cities reveal the ugly side of democracy', *Sydney Morning Herald*, 6–7 October, p. 26.
7 Australian Conservation Foundation (ACF) (2007) *Consuming Australia: Main Findings*, accessed at: http://www.acfonline.org.au/consumptionatlas, p. 10.
8 Ibid.
9 A. Moran (2006) *The Tragedy of Planning*, Institute of Public Affairs, Melbourne.
10 P.N. Troy (2003) 'Saving our cities with suburbs', *Griffith Review*, Summer, pp. 115–28.
11 H. Stretton (1975) *Ideas for Australian Cities* (2nd edn), Georgian House, Melbourne.
12 Troy, 'Saving our cities with suburbs'.
13 B. Cubby (2008) 'Our place in the sun', *Sydney Morning Herald*, 23 February.
14 A. Davison (2006) 'Stuck in a cul-de-sac? Suburban history and urban sustainability in Australia', *Urban Policy and Research*, 24(2), pp. 201–16.
15 C. Hamilton (2003) *Growth Fetish*, Allen & Unwin, Sydney.

Chapter 9

1 J. Lagorio (2007) 'U.N.'s Ban says global warming is "an emergency", posted at http://uk.reuters.com, 10 November 2007.
2 J. Diamond (2005) *Collapse: How societies choose to fail or succeed*, Viking Press, New York.
3 F. Fukuyama (1992) *The End of History and the Last Man*, Free Press, New York.

4 M. Fyfe (2009) 'Study links drought with rising emissions', *Sydney Morning Herald*, 16 August.
5 Ibid.
6 Cited in *The Australian*, Inquirer, 22–23 August, 2009, p. 26.
7 Pusey, 'In the wake of economic reform', p. 28.
8 Ibid.
9 Lovelock, *The Vanishing Face of Gaia*, p. 49.
10 C. Hamilton (2008) 'Six Degrees of Apocalypse: Recent books about climate change, *The Monthly*, April, no. 39.
11 Ibid.
12 I highly recommend one such book, Spratt, D. & Sutton, P. (2008) *Climate Code Red: The case for emergency action*, Scribe, Melbourne.
13 See http://www.ecoequity.org/.
14 Hamilton, 'Six Degrees of Apocalypse'.
15 R. Wilkinson & K. Pickett (2009) *The Spirit Level: Why more equal societies almost always do better*, Allen Lane, London.

Chapter 10

1 Lovelock, *The Vanishing Face of Gaia*, p. 22.
2 M. Davis (2007) *Planet of Slums*, Verso, London.
3 Lovelock, *The Vanishing Face of Gaia*, p. 12.
4 See http://transitiontownsaustralia. blogspot.com, but recognise that this is a very fluid webscape. Best to search for 'Transition Towns Australia' for a real-time view on this rapidly emerging network.
5 P. Mees (2000) *A Very Public Solution: Transport in the dispersed city*, Melbourne University Press, Melbourne. See also his revised and republished 2009 account, *Transport for Suburbia: Beyond the automobile age*, Earthscan, London.
6 J. Hewett (2009) 'Investors back $200bn projects', *The Australian*, Weekend Business, 8–9 August, p. 27.
7 S. Beder (2009) 'Electricity generation, climate change and privatisation', *Australian Options*, Winter, no. 57, p. 18.
8 Ibid.
9 Ibid., p. 20.
10 R. Evans (2007) *A History of Queensland*, Cambridge University Press, Melbourne, p. 196.

11 Lovelock, *The Vanishing Face of Gaia*, pp. 59–60.

Chapter 12

1 Fromm, *The Fear of Freedom*, p. ix.
2 H. Marcuse (1964) *One-Dimensional Man*, Beacon, Boston.
3 M. Pusey (2003) *The Experience of Middle Australia: The dark side of economic reform*, Cambridge University Press, Melbourne.
4 M. Latham (2009) 'Left is living in dreamland', *The Australian Financial Review*, 3 September, p. 62.
5 Fromm, *The Fear of Freedom*, p. 10.
6 Ibid., p. 14n.
7 Ibid., p. 18.
8 Ibid.

Chapter 13

1 T. Eagleton (2007) *The Meaning of Life*, Oxford University Press, Oxford, p. 155.
2 Letter to Editor, *The Monthly*, October 2009, p. 72.
3 Fromm, *The Fear of Freedom*, p. 30.
4 Ibid., p. 102.
5 Eagleton, *The Meaning of Life*, p. 147.
6 M. Pipher (1996) *The Shelter of Each Other: Rebuilding our families to enrich our lives*, G.P. Putnam's & Sons, New York.
7 R. Galbally (2009) Address to National Press Club of Australia, Canberra, 7 October.
8 Fromm, *The Fear of Freedom*, p. 69.
9 Galbally, Address to National Press Club of Australia.

Chapter 14

1 J. Schultz (2008) 'The year cities ate the world', *Griffith Review*, Winter, p. 10.
2 Lovelock, *The Vanishing Face of Gaia*, p. 157.
3 J. Gottman (1961) *Megalopolis: The urbanized northeastern seaboard of the United States*, The Twentieth Century Fund, New York, p. 79.
4 Ibid.
5 Ibid.
6 Ibid.
7 *De Civitate Dei contra Paganos* ('The City of God against the Pagans').
8 Stretton, *Australia Fair*.

SELECT BIBLIOGRAPHY

Dodson, J. & Sipe, N. (2008) *Shocking the Suburbs: Oil vulnerability in the Australian city*, UNSW Press, Sydney.

Eagleton, T. (2007) *The Meaning of Life*, Oxford University Press, Oxford.

Evans, R. (2007) *A History of Queensland*, Cambridge University Press, Melbourne.

Fincher, R. & Iveson, K. (2008) *Planning and Diversity in the City: Redistribution, recognition and encounter*, Palgrave Macmillan, Basingstoke.

Flannery, T. (2005) *The Weather Makers: The past and future impact of climate change*, Text, Melbourne.

Forster, C. (2004) *Australian Cities: Continuity and change* (3rd edn), Oxford University Press, Melbourne.

Frankel, B. (2004) *Zombies, Lilliputians & Sadists: The power of the living dead and the future of Australia*, Curtin University Books, Freemantle.

Fromm, E. [1942] (2009) *The Fear of Freedom*, Routledge, London.

Horton, S. (2005) 'The Gambler: (Re)placing the desire of money', *New Zealand Geographer*, vol. 61, issue 3, pp. 187–202.

Lovelock, J. (2009) *The Vanishing Face of Gaia: A final warning*, Penguin, London.

McManus, P. (2005) *Vortex Cities to Sustainable Cities: Australia's urban challenge*, UNSW Press, Sydney.

Marx, K. [1867] (1992) *Capital: Volume 1: A critique of political economy*, Penguin, Harmondsworth.

Mees, P. (2010) *Transport for Suburbia: Beyond the automobile age*, Earthscan, London.

Monbiot, G. (2008) *Bring on the Apocalypse: Six arguments for global justice*, Atlantic Books, London.

Self, P. (2000) *Rolling Back the Market: Economic dogma and political choice*, Macmillan, Basingstoke.

Stanley, F., Richardson, S. & Prior, M. (2005) *Children of the Lucky Country?*, Macmillan, Sydney.

Stilwell, F. (2006) *Political Economy: The contest of economic ideas* (2nd edn), Oxford University Press, Melbourne.

Stilwell, F. & Argyrous G. (2003) *Economics as a Social Science: Readings in political economy*, Pluto Press, Sydney.

Stretton, H. (2005) *Australia Fair*, UNSW Press, Sydney.

Troy, P.N. (1996) *The Perils of Urban Consolidation*, Federation Press, Sydney.

—— (2003) 'Saving our cities with suburbs', *Griffith Review*, Summer, pp. 115–28.

Wilkinson, R. & Pickett, K. (2009) *The Spirit Level: Why more equal societies almost always do better*, Allen Lane, London.

INDEX